IVAN FLORES

*Computer Consultant*
*and Professor of Statistics*
*Baruch College*
*City University of New York*

# DATA STRUCTURE AND MANAGEMENT

## Second Edition

Prentice-Hall, Inc., Englewood Cliffs, N.J. 07632

*Library of Congress Cataloging in Publication Data*

FLORES, IVAN.
  Data structure and management.

  Includes index.
  1. Data structures (Computer science)   I. Title.
QA76.9.D35F57 1977      001.6'442      76-28701
ISBN  0-13-197335-5

10  9  8  7  6  5  4  3  2  1

Printed in the United States of America

Prentice-Hall International, Inc., *London*
Prentice-Hall of Australia, Pty. Ltd., *Sydney*
Prentice-Hall of Canada, Ltd., *Toronto*
Prentice-Hall of India Private Limited, *New Delhi*
Prentice-Hall of Japan, Inc., *Tokyo*
Prentice-Hall of Southeast Asia Pte. Ltd., *Singapore*
Whitehall Books Limited, *Wellington, New Zealand*

# Contents

# Preface

This totally new edition presents the overall principles of data and their organization starting from the very beginning. The text is aimed particularly at undergraduate students in business data processing and computer or information science who have a minimum background in mathematics and computer language. Since the book builds up to a discussion of concepts of complex files and of new and novel list organization, it will also be useful for practical applications and for system programmers.

In this edition there are no embellishing mathematics or symbology. Only mnemonic operators are used and these immediately convey the action to the reader. For instance, when the *point* operator is applied to a record r in a linked list, it yields the address of the successor cell, S, which is simply conveyed by *point* r = S. The text is free of confusing typefaces. Only san serif type is used for data and list labels to make the operator notation more readable.

Since the book does not refer to programming languages or job control languages (with the exception of one or two examples), it can be used in the setting of most any computer installation. Problems are provided at the end of each chapter to help the instructor and student verify the student's accomplishments.

The first chapter develops a feeling for the nature of data. It establishes the reality from which the data are abstracted and develops units for measuring data. This chapter explains why these units differ according to where the data are found. It also presents the knowledge of the hardware and the operating system that are prerequisite to understanding how data are manipulated by the computer. Chapter 1 establishes a common background for students of different training. This chapter and the next emphasize the need to distinguish between data in the abstract and the place where the data are held momentarily. The distinction that I make, for example, between the file and the list is: the latter is a place where the file is stored.

Chapter 2 develops tools for studying data. Symbols to express relations that exist among data are presented. Graphs which convey these relations more vividly are discussed but not in a rigorous mathematical format. The aim is to provide a practical not a theoretical tool.

Assuming that we have a simple sequential file, there is still great latitude in the way that the record is structured. After establishing what it means to have a simple file structure, Chapter 3 continues by investigating the many alternatives available for setting up a record format.

The classical use of the file is the posting and updating accounting application. Files represent the earliest application of computers to the business world, and the vast majority of files are still used in this fashion. Chapter 4 describes this use to bring into play the principles already learned, so that the reader will have an early feeling for the practical application for the knowledge he is gaining.

All of Chapter 5 is devoted to the ordered list. Estimates vary about the amount of computer time devoted to sorting and ordering lists, and range from 10 to 30 % of all computer time in the business community. This chapter explains the nature of sorting and how it is used. Once having an ordered list, the binary search is an effective way for finding a particular record, and, hence, is discussed here.

Simple linked lists used only to represent linear graphs are presented extensively in Chapter 6. The use of the pointer within the record to give order to a file cannot be overemphasized. Linked lists can be extended to represent transitive graphs which are a valuable feature. Chapter 7 discusses the complications involved in extending the linked list to such structures.

The directory list is the basis of the ISAM file and represents an important technique for achieving random accessibility at low cost. Its ramifications are discussed in Chapter 8.

Mapping or hashing is the other important technique for improving accessibility. The technique is discussed in Chapter 9 where full coverage is given to searching and maintaining files under this technique.

A separate chapter is devoted to overflow which enables us to alter our files easily. Since this technique can be applied to many structures, it is discussed separately and examples are given which make the alternatives vivid.

Trees are emerging as important alternatives to structuring files for easy access. The binary tree reproduces the binary search technique without the need for calculation. It is compact and easy to use but more difficult to understand.

Chapter 11 discusses the binary tree and the need to maintain such trees in balance, along with techniques for so doing. The final section of this chapter discusses the basis for the extremely efficient VSAM file.

Having these tools, it is possible to approach complex and compound files. Chapter 12 is devoted to this topic.

# CHAPTER 1

# Data and
# the Computer

## 1.1 INTRODUCTION

### Importance

The COMPUTER is a mechanism the sole purpose of which is to work on data. The COMPUTER works extremely fast and its operation must be accurate. This can be so only if the data are organized optimally.

Since data are so important to both the user and the machine that works upon them, it is incumbent upon us to understand the alternatives for organization of the data.

### Data and the Real World

Data provide pictures or representations of the real world. The relevance and accuracy of these pictures are what make them useful to the problem solver. There are many ways that we can represent the real world. One is a photograph or sketch where we abstract the visual components. We can make other models of the world. Besides a physical model, we have learned to represent parts of the real world by mathematical formulas. These are mathematical models that, supposedly, predict results when their variables are changed according to possible events in the real world.

The computer system designer seeks to represent the real world by structuring data to represent it. **Data** are then descriptions of objects in the real

1

world. The form of this description is such that the computer can manipulate it most easily.

### Individual

Only philosophers deal with the real world in its entirety. Practical people chop off a portion of the world, which is then their province. There are only certain objects of interest for any problem solver. The collection of these objects is known as the **universe.** We shall refer to each object as an **individual.** Individuals can be of various types:

- they may be people—the people on our payroll, for instance.
- they may be objects—the parts we manufacture, for instance.
- they may be constructs—fictitious individuals—our accounts for an accounts receivable application.

It is important to realize that the universe may be different for different problems in the same company. Thus, we process the paycheck of the *weekly* employee (with a time card) in a different fashion than we do the paycheck for our executive on a *monthly* payroll. *Employees* may fall into two or more universes, even with respect to the payroll.

Each individual is different from the next, especially with regard to our particular problem. It is necessary to describe the qualities of importance to our problem for each individual. We refer to the qualities of interest as **attributes.** Each individual has attributes. It is the value of an attribute for this individual, the **attribute value**, that makes this individual unique and different from the next one.

### Descriptions

There are many attributes associated with each problem. It is easier to speak about an attribute if we name it, use an **attribute name.** For instance, in describing a person, here are three attribute names that require no elaboration: sex, height, social security number.

For each individual, each attribute has a value. For instance, for Harry for the attributes named above, we have the following attribute values: male; 5 ft 10 in.; 123–45–6789.

We have distinguished the *attribute name*—the name for a particular quality—and the *attribute value*—the value of that quality for a given individual. Each individual has many attributes by which he may be described.

Only those attributes that have relevance to the problem need be described and present in the data for the individual.

*range*          The values of an attribute may have a **range** specified for them. In the case of height, for instance, our employees will probably not be shorter than 3 ft nor taller than 8 ft. Although this range is applicable for *our employees*, it may not always be suitable. For an elementary school, for instance, where the children might not be taller but might be shorter than this range, it is unsuitable. For the social security number, clearly the range would be from 000–00–0000 to 999–99–9999. Only two alternates for sex need be recorded, although more may exist.

The description of the individual then consists of a collection of attribute values. The attributes chosen depend on the application. Those attributes irrelevant to this application are excluded from the description.

## 1.2 USER QUANTA

Data take various forms and are divided into packages of different sizes according to which agency will act upon them. We shall examine data with respect to the following agencies:

- the user—the human who enters data into the system;
- the COMPUTER;
- the medium—the place where information is stored statically for retrieval by the COMPUTER;
- storage—the site in COMPUTER MEMORY where data is kept temporarily while they are being processed by the COMPUTER.

We examine the packages or **quanta** of information with respect to these agencies in this and other sections.

### Correspondence

In Fig. 1.2.1 we see a pictorial characterization of the real world with respect to the user. There we see the universe containing individuals, each with attributes of interest to us.

For each individual there exists a collection of information that we call the **record**. The record in turn contains fields. There is one **field** for each attribute of the individual that is required for the application. The collection of records is called the **file** and corresponds to the collection of individuals that we have labeled the universe.

**Figure 1.2.1.** The universe consists of individuals each with attributes which may have one of many attribute values.

Notice that there is a correspondence in *number* in the quanta used in the real world and for the data that represent it:

- There is one universe and one file.
- The number of individuals in the universe is the same as the number of records in the file.
- The number of attributes describing the individual is the same as the number of fields in each record.

### Record

The record consists of fields; a field corresponds to an attribute. We have a name that identifies that attribute. For each attribute there is a value that describes that quality for a particular individual.

For each field in the record, there is a **field name**. This enables us to speak about the field and identify it. Within a record for a particular individual the field has a value, the **field value**. Thus, the *content* of the field, the field value, describes the attribute for this individual at this time.

### Fields

Although field values are supposed to reflect the attribute values, they need not do so in exactly the same fashion. For instance, although the individual thinks of his age in years, our application may simply consider his age group. Thus, those between 16 and 18 fall in the first group; those between 19 and 22, in the second group; and so forth.

Another example deals with the state. Although we name the fifty states, a number could just as easily represent this information. Thus New York might be 25; New Jersey, 26; and so forth.

The field value thus reflects in *some* way the attribute value for the individual. An effective data processing system will provide a current representation of the individual by recording the latest value of each attribute.

### Characters

The user atom is usually the **character**. Thus identifiers such
as names consist of strings of characters. Sometimes identifiers, such as
social security numbers, may consist of numerals only. Even so, they are
often represented in the computer as strings of characters.

Sometimes smaller atoms might be desirable. For instance, one bit suffices
to represent sex; four bits may represent a decimal digit. At this point in our
investigation we assume that the user has available all the proper characters
but does not further dissect information to a bit level. He has available this
alphabet or **character set**:

- the letters of the alphabet;
- the digits 0 through 9;
- blanks;
- special characters such as punctuation.

### Changes

The data that we have on hand should reflect the current
status of the universe. Changes can take place in the following two ways:

- The universe may change—individuals may be added to or subtracted
  from the universe.
- The individual may change—the value of his attributes may alter.

*update*    When the values of the attributes of an individual change,
this must be reflected in the file. The record for that individual contains fields, each of which should record the latest value associated
with an attribute. The action of altering those fields in the record for which
the individual's attributes have changed is called **updating**. The updated file
reflects the latest characteristics of the individuals in the user's universe.

### Maintenance

The universe changes when there are births and deaths. A
**birth** occurs when a new individual enters the universe—as when a new
employee enters the firm or when the firm acquires a new account. A **death**

occurs when an individual leaves the universe, as when an employee leaves the firm or one of their accounts cancels. **File maintenance** is the action of adding or removing records from the file to reflect the revised universe.

### Other Quanta

*library*    A **library** is a collection of files. It represents several universes with respect to a particular application. For instance, if our payroll activity is different for hourly, weekly, and monthly employees, then we may wish to keep three different files, one for each type of employee. The payroll library consists of the three payroll files and represents the combined universe—all our employees.

### Subsets

A **subset** consists of *some* of the members of our original set. We always speak of proper or nondegenerate subsets. According to our definition, the set itself *might* be considered a subset; or the subset may be empty. Both of these are *degenerate* cases.

A subset is indicated with reference to a quantum by prefixing the quantum name with *sub*. Thus, a **sublibrary** is *some* of the files of our original library. A **subfile** consists of *some* of the records of a given file. Often we shall select records for individuals that have a particular characteristic and thus create a subfile with a particular purpose.

The terms *subrecord* and *subfield* have special meanings that we define at a later point.

## 1.3 HARDWARE

### Introduction

The hardware operates on data. We need to know something about the hardware so that we shall know how it works on data and what quantum of data is transmitted between different parts of the hardware.

We don't need to know the exact details of operation for each unit. During our study we should keep in mind the following points:

- the flow of information between one point and another in the equipment;
- the quantum of data moved;
- the limitations of the equipment and why they exist.

### The System

A typical computer system is composed of five subsystems, as presented in Fig. 1.3.1. Each subsystem is, for the most part, autonomous and operates according to its own rules. It is stimulated to perform its actions by signals it receives from other subsystems. It receives data and transmits them to other subsystems.

Central to the computer system is the MEMORY SUBSYSTEM, or simply, **MEMORY**, the internal MEMORY of the COMPUTER. Its purpose is to hold information that it makes quickly available in nanoseconds (billionths of a second) to other subsystems. It accepts data from other subsystems, which it memorizes also in a matter of nanoseconds.

Actually, MEMORY works at a speed in the range of 100 nanoseconds to a microsecond (millionth of a second) or two. Sometimes it is better to have information available even more quickly. A set of REGISTERS called GENERAL PURPOSE REGISTERS, or GPR's, provide data in a matter of 10–100 nanoseconds.

The **PROCESSOR** receives data that it operates upon to produce new data. A datum to be processed is called an **operand**. One or more operands is worked upon by the PROCESSOR to produce a **result**, which is cometimes called an operand too. The result is a *new* datum. The PROCESSOR is creative!

**Figure 1.3.1.** The computer system

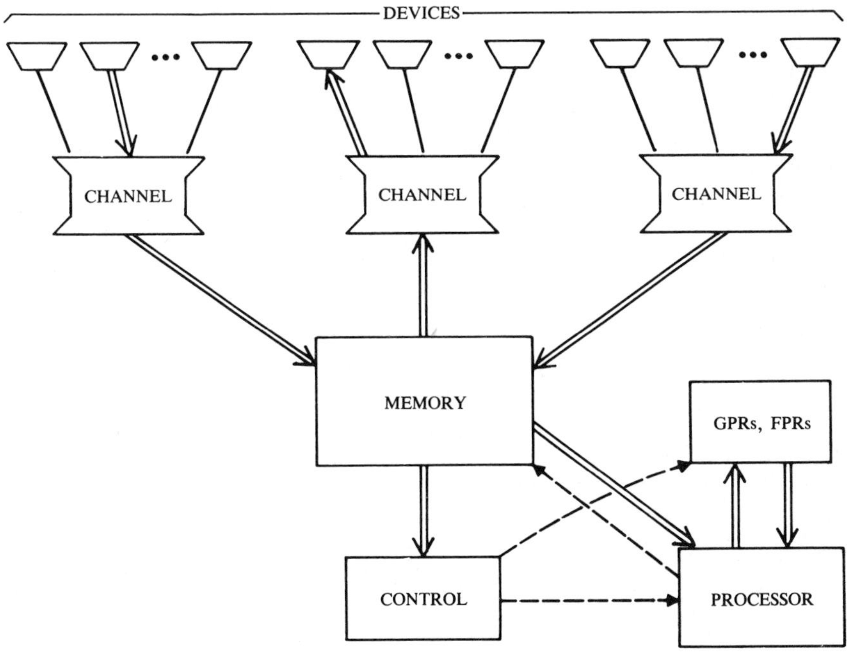

In modern COMPUTERS, information is brought into and taken out of MEMORY for the outside world by means of the CHANNEL. The CHANNEL enables the transfer of data between the DEVICE and MEMORY while the COMPUTER is processing. The ability to process and bring in and move out data is called **overlap** or **simultaneity** and is a necessity in modern COMPUTERS.

CONTROL is the subsystem that directs the COMPUTER. It fetches instructions from MEMORY, interprets them, and directs the other subsystems. In Fig. 1.3.1 we see that MEMORY provides instructions of the program to CONTROL. CONTROL does not furnish any data to MEMORY.

### Composition

At the lowest hardware level we find the **electronic atoms** that are such things as FLIPFLOPS and GATES. They needn't concern us here. These atoms are assembled to form functional units. Each FUNCTIONAL UNIT performs a function that is well-defined:

- Inputs provide data to be operated on.
- Outputs provide results.
- Control signals tell the unit what to do.

FUNCTIONAL UNITS and atoms are assembled together to make the subsystems. We have seen how the subsystems are connected to create the overall computer system.

Let us now get some feel for what the functional units are. Figure 1.3.2 shows pictorially four such units. At the top we see the **REGISTER**. Its function is to store a datum temporarily and make it available rapidly—in a few nanoseconds—to other units in the subsystem. The input datum is found at in. It can be obtained by the user at out. The request to enter a new datum appears.

Another important functional unit is the **SWITCH**. It directs the flow of data from one place to another within a subsystem. The SWITCH shown in Fig. 1.3.2 has two inputs, **in1** and **in2**. Each input can be connected to any one of the outputs, **out1**, **out2**, or **out3**. The path connecting each input to an output is determined at CONNECT.

The functional unit called the **COUNTER** contains an input at in that consists of a number of pulses that it is to count. The output of the COUNTER at out is the number of pulses that have appeared at **in** and is found there in binary form. Sometimes it is necessary to reset the COUNTER to 0, and there is an input provided for this purpose.

The function of the **ADDER** is to add two numbers presented to it in coded form and it provides the sum in coded form as shown pictorially in Fig. 1.3.2 also.

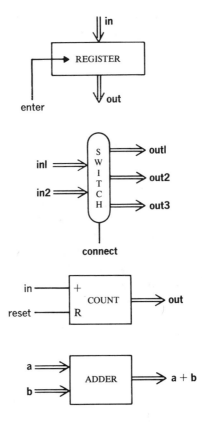

**Figure 1.3.2.** FUNCTIONAL UNITS of the COMPUTER

### Memory

The layout of MEMORY is presented in Fig. 1.3.3. MEMORY consists of a number of **word cells**, each of which can hold a datum. The size of each word cell is defined by the hardware design. A word is rated in bits. The word size of MEMORIES of current COMPUTERS varies from 8 bits to 72 bits.

Associated with each word cell is an **address**. This enables us to pinpoint a particular word cell in MEMORY that is to receive or produce a datum.

MEMORY performs three functions:

- It *holds* information, preserving it for future use.
- It *recalls* information requested by another subsystem.
- It *memorizes* information, placing a datum in a specified cell.

***Holding*** information is a static function and requires no further description.

***Recall*** is nondestructive. MEMORY makes a copy of a datum and provides it to the user. The original datum still resides in the word cell from which it was copied.

***Memorization*** places a datum at an address. Both the datum and the address are supplied by a user. ***Memorization*** is destructive. The datum previously held at the address at the time of the request is destroyed and is overwritten with a new datum.

*functional*    MEMORY consists of four functional units. These units are
*units*    shown in Fig. 1.3.3:

- MEMORY ADDRESS REGISTER, MAR, holds the address provided by the user.
- MEMORY DATA REGISTER, MDR, holds the datum for memorization or receives the datum during recall.
- The MEMORY CONTROL UNIT, MCU, receives the signals indicating the action the MEMORY is to fulfill and produces a signal done, indicating the MEMORY's action is finished.
- The word cells hold the data.

**Figure 1.3.3.** MEMORY

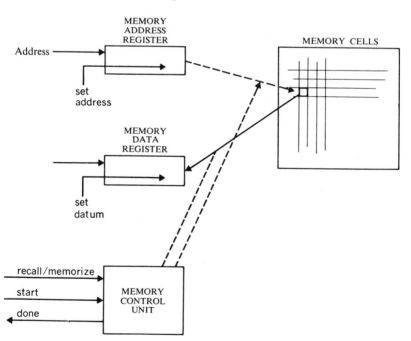

*recall*   *Recall* is to retrieve a word from MEMORY. To do so, we

1. supply an address to MAR,
2. supply a signal recall to the MCU,
3. wait till the signal done is produced by the MCU,
4. take the word from the MDR.

*memorize*   *Memorize* is to place a word in MEMORY. To accomplish this, we

1. supply an address to MAR,
2. supply a datum to MDR,
3. supply a signal memorize to MCU,
4. refrain from using until done is produced by MCU.

**Processor**

The PROCESSOR in Fig. 1.3.4 contains

- an ADDER and COMPLEMENTOR to do arithmetic;
- REGISTERS to hold operands (quantities to be operated on);

**Figure 1.3.4.** PROCESSOR

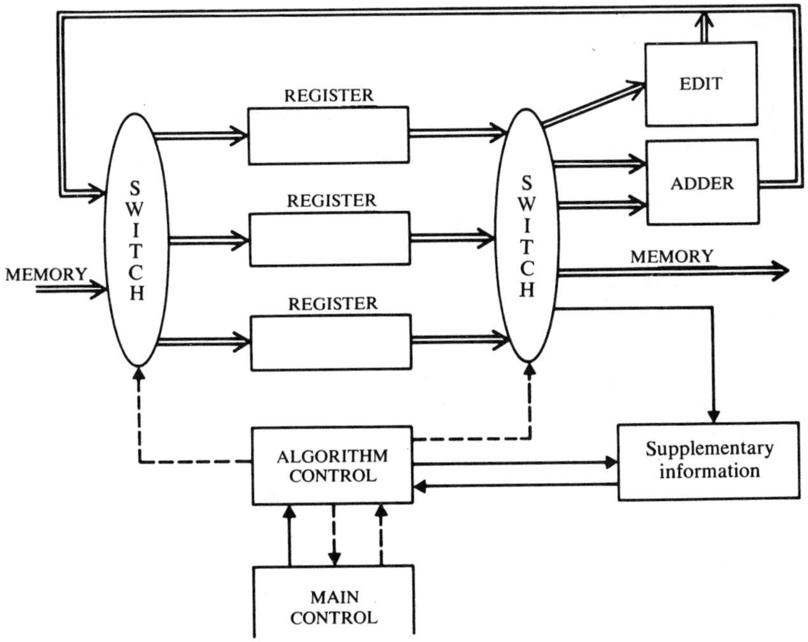

- SWITCHES to control data flow;
- COUNTERS and FLIPFLOPS to keep track of the present state of processing;
- an EDIT UNIT for bit operations;
- a CONTROL UNIT, the PCU, to supervise (hardware) what goes on.

The PROCESSOR can do

- arithmetic: $+$, $-$, $\times$, $\div$;
- editing: shifting, extraction, masking, and logic.

**Control**

CONTROL in Fig. 1.3.5 contains

- an INSTRUCTION ADDRESS REGISTER, IAR—depending on the machine, it contains the address of the current or next instruction (System/370);
- an INSTRUCTION REGISTER, IR, which holds the instruction to be executed;
- INSTRUCTION DECODER, ID, which interprets the instruction;
- an INSTRUCTION CONTROL UNIT, ICU, which delegates the instruction to the proper subsystem and controls that flow among subsystems.

*fetch*        After a command is finished, the process of obtaining the next command is called **fetch**:

- IAR now contains the address of the next command.
- Send it to MAR.
- The MEMORY brings a datum to the MDR.
- This goes to the IR.

*execute*        After fetch comes a cycle called **execute**:

- The instruction to execute is in IR.
- ID sends an interpretation to ICU.
- ICU may request address calculation.
- The PROCESSOR (or CONTROL ARITHMETIC if present separately) calculates an effective address, if needed.
- The MAR is sent this address.
- MEMORY obtains a datum.
- The operand is routed from MDR to the DESTINATION REGISTER by the ICU.
- When processing is required, it is done by the PROCESSOR.

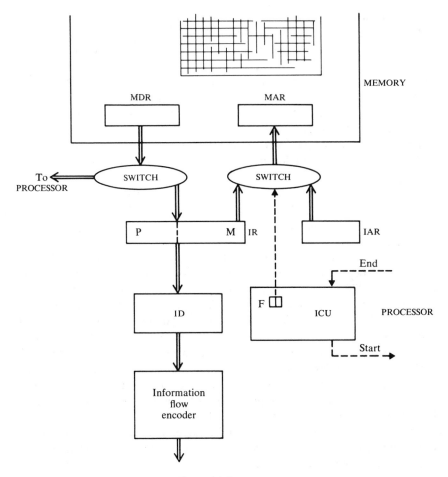

**Figure 1.3.5.** CONTROL

This description applies to conventional arithmetic and editing; branching and so forth are described elsewhere.

## 1.4 COMPUTER QUANTA

### Bit

The smallest unit of information in the COMPUTER is the **bit**. This is a contraction of binary digit. A binary digit is one for which there are only two alternatives. These are signified as 1/0, on/off, or yes/no.

The binary system was chosen for COMPUTERS early in the game. In the 1940's it was easy to make bistable devices—those that have two stable states —which were reliable and less expensive than devices that have several stable states. As a consequence, since such devices were readily available and inexpensive, COMPUTER manufacturers used them throughout their equipment.

Since COMPUTERS are composed of bistable devices, this dictates the property of data and its representation in the COMPUTER. Thus, regardless of the symbol to be represented, its representation consists of an array of bits; the number of bits to be used depends on the extensiveness of the alphabet to be used.

### Character

The alphabet—the characters, symbols, and digits available to the user—depends on the range of applications to be made available to the user. System/370 uses the Extended Binary Coded Decimal Interchange Code, EBCDIC, which is an eight-bit code providing a possible alphabet of 256 symbols. Since most print chains contain only 48 different symbols, this seems rather wasteful. This means that each character, regardless of how it will be printed, is represented in the COMPUTER as an array of eight bits, a **byte**.

Other computers have seen fit to use codes providing only six bits per character. They seem to be able to provide the same services. Since this book uses System/370 as its primary example of a COMPUTER, however, discussions that follow refer to the byte rather than the character.

For System/370, MEMORY is **byte addressable**. This means that regardless of the MEMORY word size, some commands can pinpoint fields by addressing the exact byte that begins that field.

### Digits

Some machines provide decimal arithmetic. Naturally the operands that participate are decimal numbers. Each digit of a decimal number requires only four bits for representation. If numbers were left in the eight-bit character representation, their storage in the COMPUTER and outside the COMPUTER between calculations would be less efficient.

Numerical information is brought into the COMPUTER from an original document in character or eight-bit format. The action that changes it from eight bits per character to four bits per digit is called **packing**, and decimal information is said to be in a **packed format** in this case.

### Binary

Some COMPUTERS, including System/370, provide binary arithmetic besides decimal arithmetic. The operands for binary arithmetic must be binary numbers. Such numbers are 32 bits long and are called Words by IBM. Notice the initial capital for Word to distinguish IBM's definition from the other definition (the quantity of information in a MEMORY word cell). A Word is then a number represented in the binary counting system where negative numbers are represented in two's complement format.†

System/370 also provides the binary halfWord. This is a 16-bit two's complement number and is used for small quantities.

### Hexadecimal

The System/370 code for digits, the EBCDIC code, is an eight-bit code that does not yield easily to simple description in a verbal discussion. To describe these combinations, users have broken the byte into two parts, each of which is called a **nibble**. Each four-bit nibble has sixteen possibilities. In **hexadecimal**, the combinations 0000 through 1001 designate the decimal digits, represented in the binary number system (base 2). Thus 0000 is zero and 1001 is decimal 9. This leaves six combinations 1010 through 1111 unaccounted for. These are assigned the symbols A through F, respectively.

To see an example of this, the letter A (byte) is represented in hexadecimal (hex) as C1, which in turn corresponds to the eight-bit combination 11000001. The left nibble, 1100, is arbitrarily called C; the right nibble is recognized as 1 in both binary and decimal. Thus

$$A \equiv C1 \equiv 11000001 \tag{1.4.1}$$

where $\equiv$ means *represented as*.

B is represented as hex C2 or binary 11000010.

### Information Flow in the Computer

In Fig. 1.4.1 we see the subsystems of the COMPUTER with an indication of the quantum of data that flows between each. Its importance is to understand how much information is exchanged and why.

---

† Two's complement format is described in more detail in Ivan Flores, *The Logic of Computer Arithmetic*, Prentice-Hall, Englewood Cliffs, N.J., 1965.

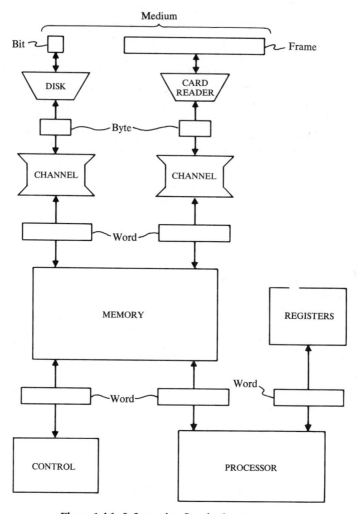

**Figure 1.4.1.** Information flow in the COMPUTER

## 1.5 MEDIA QUANTA

### Introduction

Data are frequently stored on **intermediate media,** storage, from which data are available only when mounted on an IO DEVICE. The medium *is* the physical configuration that stores information. It is much easier and quicker for the COMPUTER to recover data from such media. The speed factor is as much as five orders of magnitude greater than for a direct entry, a KEYBOARD. The following gives you some idea of this:

- The human can enter information via a KEYBOARD at about 15 strokes per second.
- A punchcard can be read at about 2000 characters per second.
- It is feasible to read in from disk or drum at 2 million characters per second.

The most important fast intermediate medium is the disk. The punchcard, also an intermediate medium, is much slower to use.

A **medium** is the physical configuration that stores information. The COMPUTER operates a DEVICE that either

- enters information into the medium or **writes**;
- extracts information from the medium or **reads**.

The DEVICE for performing reading or writing is described next. All fast DEVICES (such as the DISK DRIVE) require that the medium (the disk) be moved while reading or writing is performed. A direction of motion is thus defined with respect to the medium—the longitudinal direction.

### Device Components

Figure 1.5.1 shows the functional components that exist in most DEVICES.

**Figure 1.5.1.** DEVICE components and action

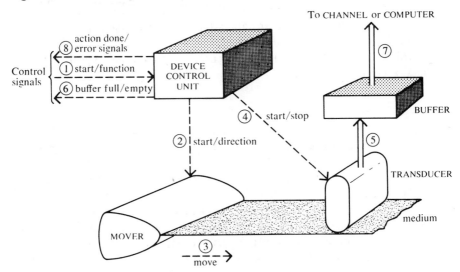

A TRANSDUCER detects the data on the medium and converts them into impulses that can be used by the COMPUTER. Alternatively, the same or a different TRANSDUCER may be used to take signals produced by the COMPUTER and cause them to be recorded upon the medium.

There is a MOVER that positions the medium so that the proper segment of the medium is beneath the TRANSDUCER. In some cases the signal conversion is done while the medium is moving. Hence it is the task of the MOVER not only to position the medium but to keep it going at the proper speed.

A BUFFER is the hardware element in the DEVICE for accumulating data. This is generally necessary because the unit quantity of data handled by the TRANSDUCER is different from the unit quantity accepted or received by the DEVICE proper.

Finally, there is a DEVICE CONTROL UNIT—DCU—whose purpose is to receive and interpret signals from the COMPUTER (and the human). It then relays these singals to the MOVER, TRANSDUCER, and BUFFER, directing them how to perform.

### Transducer

The purpose of the TRANSDUCER is to pass data between the COMPUTER and the medium. The COMPUTER handles data in the form of electric impulses; data are stored on the medium in the form of a physical alteration of the medium. Conversion between these two data forms is performed by the TRANSDUCER: for input, electrical signals cause the medium to deform; for output, the deformation is detected and electrical signals produced. In some cases (the HEAD in the DISK DRIVE) a single TRANSDUCER can move and convert data in both directions (bidirectional); in other cases one TRANSDUCER is needed to enter data, while a separate one is required to retrieve it (unidirectional). Thus, a card punch enters holes into a punchcard; a card reader detects the holes in the card and provides electrical signals to the COMPUTER.

### Medium

Most intermediate storage media are flat or can be (hypothetically) cut so that they can be stretched out onto a flat surface. In any case, they are surfaces. We can superimpose two dimensions at right angles to each other on each surface, as shown in Fig. 1.5.2. One of these dimensions is in the direction of movement to which the medium is subjected. This applies whether the medium is given rotational or translational motion (straight-line movement). Translational motion occurs for a punchcard as it passes through the punchcard reader.

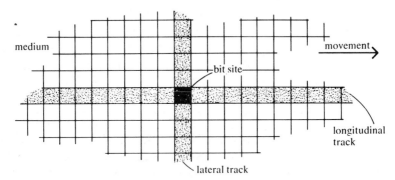

**Figure 1.5.2.** Dividing the medium into tracks and bit sites

In Fig. 1.5.2 we see a grid superimposed upon the surface. The direction of motion is called the **longitudinal direction**; perpendicular to this is the **lateral direction**. The TRANSDUCERS lie in a plane in the lateral direction. Each TRANSDUCER traces out a **track** in the plane in the longitudinal direction.

### Frame

We define a **block** as the amount of information that can be read from the medium by the DEVICE at a single request from the COMPUTER. Let us now take the grid description of the medium given above and fence off on it the portion that comprises a block. As shown in Fig. 1.5.3, the block consists of portions of **longitudinal tracks** and **lateral frames**. A frame is lateral and is the area subtended by the set of TRANSDUCERS. Thus, in the figure, you see a number of longitudinal tracks and lateral frames. The total area is bounded by a rectangle—this is the block. Further, the intersection of a longitudianl track and a lateral frame is a **bit site** where a single bit of information can be recorded.

**Figure 1.5.3.** The block and its frames

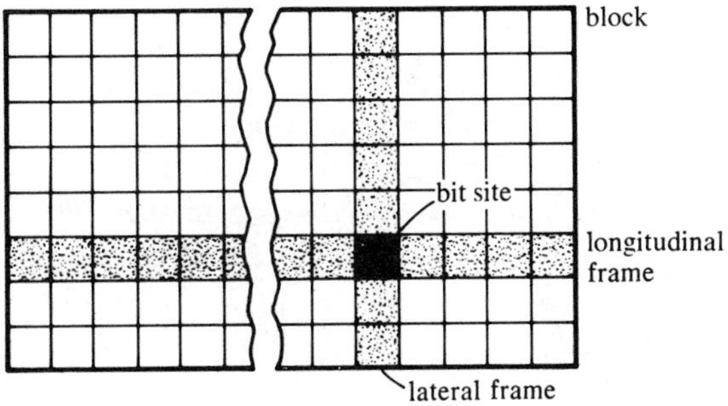

In sections that follow we shall delineate the structure of the block for various media. It is important to distinguish the number of frames that make up the block in each direction. In some cases the number of lateral frames in a block is variable because the medium permits variable block size.

### Quanta

*block*  The block is the amount of information that can be read from the medium by the DEVICE to the COMPUTER in a single request. The block is also the quantity of information that the COMPUTER provides to the DEVICE to place on the medium with a single request.

Whose request? We could consider the user's request, but this would involve us in user quanta. It is important to make the level of the request very specific. The user request actually goes to software that activates the COMPUTER. It is this request from the software that is given at DEVICE level and is given by the COMPUTER or his agent. When the COMPUTER has a CHANNEL, it is the CHANNEL that issues the request to the DEVICE.

### Volume

How much of the medium can be addressed by the COMPUTER at any given time? Certainly no more than is available upon the DEVICE. The amount of the medium that is mounted upon the DEVICE is called the **volume** and provides the limiting factor for accessibility.

*examples*  For the TAPE DRIVE the *volume* is a **reel**. Figure 1.5.4 shows a reel of tape. Upon it we note a block. The block consists of a number of frames; each frame for tape is a character.

In Fig. 1.5.5 we see a punchcard that is part of a deck. A **deck** is a volume of punchcards. The block is the punchcard itself. The (longitudinal) frame for a punchcard is a **column**; each column represents a character. The track is a **row**.

**Figure 1.5.4.** Tape quanta

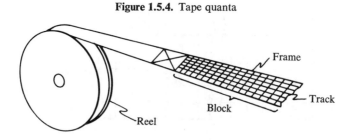

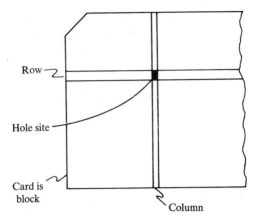

**Figure 1.5.5.** Card quanta

In Fig. 1.5.6 we see part of a **disk pack** that is the volume for a DISK DRIVE. Each of the disks is divided into tracks that are longitudinal frames. Each track is further divided into blocks. A block for a disk is hence a track segment. The track consists of lateral frames that are also bit sites since the track is recorded serially by bit. Hence, the character (byte) consists of (eight) bits along the track, as shown in the figure.

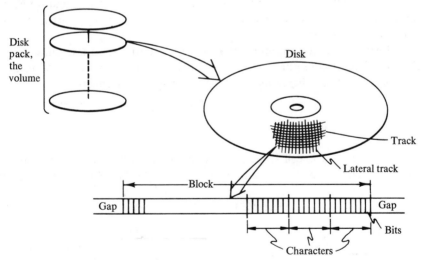

**Figure 1.5.6.** Disk quanta

### Block Sizes

For some media, such as the punchcard, the block size is fixed. The user has no control over it—that's the way the card comes.

For other media the user does have control of the block size and, within limits, when he creates his file, he can fix the size of blocks on his medium. Now, even though this size is variable at the inception of the file, since he does not alter it thereafter, this file is said to have **fixed block size.**

Further latitude can be provided with some media where the user merely indicates the maximum block size he needs. Then, for his application, blocks may vary in size as long as they do not exceed this maximum. This is called **variable block size.**

## 1.6 STORAGE QUANTA

### Introduction

We have examined data with regard to the user, to the COMPUTER, and to the medium upon which it is stored. We still lack a proper means for discussing files while they are in the act of being processed. At this point the files exist either in COMPUTER MEMORY or out on some auxiliary storage medium where they are fairly readily accessible. We shall discuss how parts of files are moved around from one place to another, and we need terms to discuss the place where information is stored momentarily as it is being moved around.

Unfortunately, the literature does not have terms for these sites. **Cell, list,** and **element** are terms that are redefined here to have very special meanings and will be used hereafter only with those special meanings.

### Cell

The **cell** is the place where a record is stored. In Fig. 1.6.1 we see a record consisting of a number of fields. At the right we see an area

**Figure 1.6.1.** The record and cell

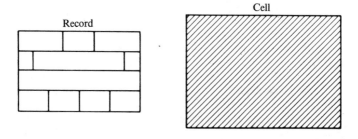

of MEMORY just large enough to accommodate the record; it is this area that is called a cell. Since the cell is part of MEMORY, it has an address. We may give the cell a symbolic name such as MOICELL. MOICELL then is the name applied to the cell, and it is also the address of the first unit of information comprising the cell. Thus, for System/370, MOICELL is the address of the first byte of this record area.

In COMPUTERS where information is allocated on a word basis, a cell should begin on a word boundary so that it may be addressable, and so forth.

Figure 1.6.2 shows that when the record is moved into its cell, it just fits snugly. Note carefully that if records in my file vary in size, then the cells that contain them must also vary in size, and one cell can properly hold only some, not all, of the records in my file. For the moment, we restrict our discussion to files that have records all of the same size (**fixed-size records**).

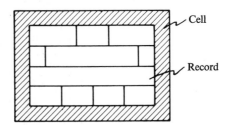

**Figure 1.6.2.** The record in its cell

### List

A **list** is an area in storage that holds a file. Since there must be room in the list for each record in the file, we can say that the number of cells in the list must equal or exceed the number of records in the file. An example is shown in Fig. 1.6.3. At the time that we process a file, it resides in a list. Some cells or areas in the list may be empty. This is an important consideration, for if some areas are empty, how do we recognize this? If an area contains garbage, how can we distinguish garbage from the real thing?

In Fig. 1.6.3 we see that our list is packed loosely; there are holes—empty areas—that contain no records.

### Element

The list holds the file; the cell holds the record. What holds the field? An **element** is the area in MEMORY where the field resides.

List

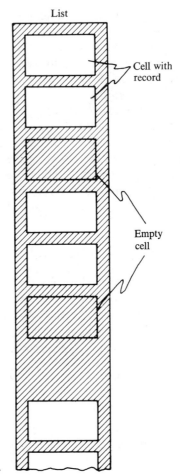

Cell with
record

Empty
cell

**Figure 1.6.3.** A file resides in a list.

## 1.7 INPUT/OUTPUT SOFTWARE

### Need

There are several reasons why we have elaborate software to perform the input and output functions for the user. These may be summarized as follows:

- quanta conversion to simplify the user's request format;
- anticipation of the user's needs to save the user's time;
- coordination of many users for many DEVICES;
- conversion of user symbolic request to a physical medium location.

*quanta*    When the user needs information, it is generally in the quantum of one new record. Once he has this record, his program can address the various fields that comprise this record. It is much more economical, however, to store data in blocks, each of which contains several records. The software makes this blocking transparent to the user, servicing his requests in terms of records but in turn making physical requests of the hardware with respect to blocks.

*timing*    When the user needs a new record, he asks for it. Depending on the file structure, the software may request the block containing the record *before* or *after* the user request is received. If the file structure does not convey the nature of the next user request to the software, the physical request for a block cannot be performed until the user request is given. The record can become available only after the user has designated it.

In contradistinction, when processing is serial in nature, and the software knows which record the user will request next, the software can anticipate it and have the record ready before the user has asked for it.

*coordination*    Our large COMPUTERS permit **multiprogramming**, several users, each with their programs in MEMORY (sometimes called CORE) at the same time. Each user has his own files, stored on media on different DEVICES. Actually, different users may store different files on the *same* medium. It is even possible for different users to access the *same* file. Obviously, a means is required to control these users so they will not mess up one another.

### Two Components

Modern input/output software consists of two components:

- The first services the user's needs.
- The second operates at the hardware level and actuates it.

The names of these components differ from one manufacturer to another and even differ within the software provided by the same manufacturer.

In OS and VS for System/370, the user speaks to an **Access Method (AM)**. For DOS for System/360, this is called the Logical Input/Output Control System (LIOCS). Other manufacturers have different names for this user IO processor. We shall speak of AM's

In OS and VS for System/370, the **Input/Output Supervisor (IOS)** communicates directly with the hardware and coordinates DEVICE actuation. For DOS this supervisor is called Physical IOCS, or PIOCS.

Hereafter we shall only use the System/370 VS terms but the concepts are universal nonetheless.

### IOS

The purpose of IOS is to coordinate COMPUTER-wide DEVICE activation. There must be a single agency to do this. Hence IOS acts as a clearing house and monitors what every DEVICE is doing.

All requests for input or output are presented to IOS. If he cannot service the request immediately, he records it and places it into the queue where the request will await its turn for service.

When a request is ready for service, it is IOS that issues the hardware command (SIO for System/370) to perform the required activity.

When the DEVICE has completed its action, it is IOS who determines the state of termination. That is, the activity

- could complete properly;
- might terminate with an error;
- might come up with an exception condition, as when the file is exhausted.

Again, it is IOS who is responsible for recovery tactics. Thus, if an error has been detected, he may be able to reread the information and come up with an error-free block.

### Access Method

Each file that may be required by the program must have an Access Method established for it before data in the file can be accessed. Both higher languages and assembly language provide a verb such as OPEN to communicate to the Operating System that the Access Method should be established.

Under multiprogramming, for each file that each user requires there is one Access Method present in CORE. For a large installation this represents many Access Methods.

The program directs requests to the Access Method for new records or asks that records it has created be written out. Thus the user converses in terms of records. But the Access Method collects records into blocks as required by the medium and the DEVICE or divides blocks into records for the user. This is called, respectively, **blocking** and **deblocking**. The Access Method or AM provides space to hold a block temporarily while it is being used or constructed. This space is called a (software) **buffer**. Along with each buffer is a **channel program**—instructions to the DEVICE on how to handle information and the buffer. Finally, there is a control block for each buffer—an **input/output block** or **IOB**—which monitors the progress of activity with respect to the buffer. When the AM is established, enough buffers, channel programs, and **IOB**'s are constructed to satisfy the user.

### Interrelation

In Fig. 1.7.1 we see the interrelation among the two IO components, the program and the hardware. The program makes its request of the AM using one of the four verbs GET, PUT, READ, or WRITE. The AM will supply the record or accept the record from the program and try to perform the action required. In so doing, the AM may require input or output with respect to one of the blocks with which it is working. To make such a request, the AM issues a request to IOS using the verb EXCP for *EXecute Channel Program.* The IO Supervisor may not be able to satisfy this request immediately if he has a backlog of other requests from other users still to be serviced.

When IOS is ready to initiate IO for this request, he issues the hardware command SIO that activates the CHANNEL and the appropriate DEVICE to

**Figure 1.7.1.** Interrelation of IO software.

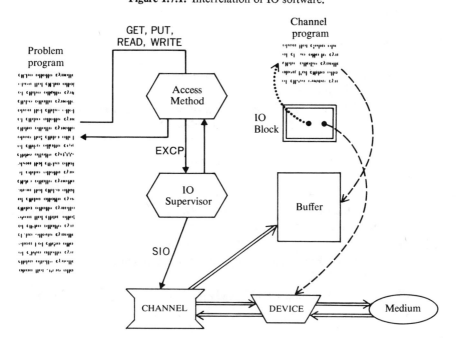

perform the action. The CHANNEL sends hardware signals to the DEVICE to start it; when the DEVICE is done, it sends a signal to the CHANNEL.

The means for communication between the AM and IOS is a set of control blocks available to both, of which the **IOB** is an example.

## PROBLEMS

1.1   Consider a file to keep an inventory of all the products your firm makes.
(a) What is the universe?
(b) What is the individual?
(c) What attributes are of value for the application?

1.2   For an attribute distinguish between these:
(a) the attribute name;
(b) its value;
(c) the value range.
Why are these to be considered in designing a field?

1.3   Given a field of $n$ bits, what is the *maximum range* of values that can be conveyed by it?

1.4   For Problem 1.3, how is the range affected by the kind of characters in which the value is encoded, namely,
(a) binary,
(b) decimal,
(c) character,
(d) hexadecimal.

1.5   Distinguish between the *cell* in the *list* and the *word cell* in MEMORY. What determines the size (in bits) of each?

1.6   How does the programmer *keep track of* the cell containing his record? the word cell of interest? Explain in terms of a programming language such as COBOL and BAL.

1.7   Explain why, when MEMORY is accessed,
(a) for *recall*, old data is preserved;
(b) for *memorizing*, old data is destroyed.

1.8   How does the COMPUTER distinguish between a datum and a command for a given word cell in MEMORY?

1.9   What is *byte addressability*? What advantage does it provide to the business user? Explain carefully. What does it cost in terms of the COMPUTER design?

1.10  Explain the advantages and disadvantages of an eight-bit character code.

1.11 What is *direct entry*? Contrast it with the intermediate medium. Why is the latter so prevalent?

1.12 What is the necessity of the MOVER in a PERIPHERAL DEVICE? Describe the needs of the MOVER with respect to
(a) speed,
(b) accuracy,
(c) consistency.

1.13 Distinguish among the following:
(a) track,
(b) frame,
(c) block,
(d) bit site,
(e) lateral,
(f) longitudinal.

1.14 Give examples of volumes and blocks for various media:
(a) punch cards,
(b) magnetic tape,
(c) magnetic drums,
(d) disks.

1.15 (a) What is variable about the fixed block size?
(b) What is fixed about variable block size?
(c) Why?

1.16 Define list *cell, list,* and *element* as used in this book. Why are such terms needed?

1.17 In general, what is the operating system, and why is it needed? Can a COMPUTER function without it? Why are some operating systems so big?

1.18 Explain the need and functions of the Input/Output Supervisor and the Access Method. Are both always necessary? Explain.

1.19 What is blocking and deblocking? Why is this often left to the AM? What are the advantages of blocking records?

1.20 How do the AM and IOS communicate? Why is it that only the IOS can issue EXCP?

# CHAPTER 2

# *Tools for*
# *Studying Data*

## 2.1 SYMBOLS

### Need

In the discussions that follow we shall have to refer to specific and relative members of both lists and files. The more ways that we can provide to help us distinguish between files and lists and between records and cells the easier it will be. To help us to this end, capital or uppercase letters are used for storage quanta and small or lowercase letters, for user quanta.

Next we would like some way to distinguish larger from smaller entities. Again, the technique we use is arbitrary. Boldface type represents the largest entities. Thus, symbols for files and lists will be in boldface. Medium-sized entities such as records and cells use symbols in the standard typeface. The finer subdivisions of each appear in italics. Thus, the symbols for fields or elements will appear in italics. Table 2.1.1 is an example that should make the use of these conventions clear.

### Position Relations

We frequently want to show what is happening with cells and records of interest as processing takes place. There are shorthand ways of presenting this that are effective.

**Table 2.1.1.** Quanta and Type Conventions

| Description (size) | Reality | Abstraction | | Realization | |
|---|---|---|---|---|---|
| | | Data | Symbol | Storage Data | Symbol |
| Largest | Universe | Library | | Data base | |
| | Application subset | File | a | List | A |
| | Individual | Record | $a_i$ | Cell | $A_i$ |
| Smallest | Attribute name value | Field name value | $a_{ij}$ | Element | $A_{ij}$ |

***cell contents***    The cell holds a record. To show which record is contained in a particular cell, parentheses act as an operator to withdraw the record from the cell. Thus, to say that record a is contained in cell A, write

$$(A) = a \qquad (2.1.1)$$

***record***    The inverse procedure is, given the record, find the cell
***location***    that now holds that record. Brackets perform this operation. Thus, to show that the record a is located at cell A, write

$$[a] = A \qquad (2.1.2)$$

***transfer***    That data are moved from one place to another is indicated symbolically with an arrow: the tail of the arrow is *preceded* by a datum—the one that is being moved; the head of the arrow is *followed* by a location—the place where the datum is put. To show that record b is moved to cell D, write

$$b \rightarrow D \qquad (2.1.3)$$

To move the record found in cell E to cell F, write

$$(E) \rightarrow F \qquad (2.1.4)$$

To show that record g will replace record h at the cell now occupied by h, write

$$g \rightarrow [h] \qquad (2.1.5)$$

### Key

Each record represents an individual. The individual is recognizable in his universe by his characteristics and hopefully can be distinguished from other members of his universe. Similarly, we need a way to distinguish one record from another and, further, absolutely to identify a record with the individual it represents.

To identify a record there is one (or more) field(s) in that record called the identifier or **key field.** When more than one field is used to identify a record, this is considered a special case. The several fields that make up the identifier could be regrouped into a single field and thus the special case reduced to the general case.

Henceforth instead of speaking of a key field, we simply call it a **key.** The key of a record is thus the value of the key field and uniquely identifies a record.

We shall use the letter K as a subscript to indicate the key field or, for the record or element in the cell, where it is located. Thus, $a_{ik}$ is the value of the key field for record $a_i$. It is found in cell $A = [a_i]$ in the position $A_{iK} = [a_{iK}]$.

### Order Relations

Order, one of the most important properties of a file or list, is defined using the key of each record. To examine this property further, break down the key value into the characters that convey it. For these characters, which are atoms of representation, we define an **order relation** (or, sometimes, **collating sequence**). This same order relation must be perpetuated in the code used to represent information in the COMPUTER.

For instance, we may set up the following inequality that exists among the characters in the alphabet for our application:

$$A < B < C < \cdots < Z < 0 < 1 \cdots \text{ etc.} \tag{2.1.6}$$

To make the COMPUTER perform properly in our service, the code for the character A should be less than the code for B, etc. This is true for the EBCDIC code where the hexadecimal combination C1 represents A, C2 represents B, and so forth. The combinations C1 and C2 are binary numbers in the COMPUTER. The binary number representing A is smaller than that representing B, so that we have

$$A \equiv C1; \quad B \equiv C2; \quad C1 < C2$$

where $\equiv$ means *represents*.

$$\tag{2.1.7}$$

The decimal digits should have codes that are larger than the letters to satisfy (2.1.6). We note that Z is represented by the hexadecimal E1, whereas 1 is represented by the hexadecimal F1, so that we have

$$Z \equiv E1; \quad 1 \equiv F1; \quad E1 < F1 \qquad (2.1.8)$$

The key field is made up of several characters. When the field includes a blank, the *blank* has a code of 40. For the moment, let us consider keys as fields of fixed length. To rank two records, compare the keys for the records. We now have an ordering relationship that yields, for instance,

$$DAVEb < DAVID;$$
$$C4C1E5C540 < C4C1E5C9C4 \qquad (2.1.9)$$

where $b$ above represents the blank.

### Record Rank

It would not be proper to say that one record is smaller than another if we are comparing the keys, not the length, of those records. We shall, however, use this shorthand: we shall say that the **rank** of a record $a_i$ is lower than the rank of another record $a_j$ if (the value of) the key of the first record is smaller than (the value of) the key of the second record. Thus,

$$a_i < a_j \equiv (a_{iK}) < (a_{jK}) \qquad (2.1.10)$$

### Cell Rank

The cell is designated by an address. The address either is a binary number or is represented by a combination of characters or digits that can be converted into a binary number. Thus, the address itself is a means for ranking the cells where data are found. The subscripts that we apply to cells are numerals and a cell with a lower numbered subscript indicates a lower number address, one that comes first in MEMORY. Thus $A_{31}$ is before $A_{42}$ and, in general,

$$i < j \Rightarrow A_i < A_j; \quad A_{31} < A_{42} \qquad (2.1.11)$$

where $\Rightarrow$ means *implies*.

Records reside in cells of our list. The position of a record in the list is important to us and we should know, in traversing the cells from low address to high, whether we encounter one record before or after another. The first record that we encounter is said to precede one that we encounter later, and we use precede operator $\prec$ to designate this. We can say this symbolically thus:

$$a_i \prec a_j \Rightarrow [a_i] < [a_j] \qquad (2.1.12)$$

Translated into English, (2.1.12) reads thus: "If the record $a_i$ precedes the record $a_j$ in some list, then the cell occupied by $a_i$ has a lower address than the cell occupied by $a_j$."

## 2.2 RELATIONS AND GRAPHS

### Relations

We shall constantly be dealing with relations among data, cells, and other variables. In this section we examine pictorial ways of presenting these relations. Right now, let us investigate the various kinds of relations that exist; and before doing this, let us provide a temporary means for representing a relation symbolically.

A relation exists between two objects or individuals. Suppose we represent these individuals by lowercase letters, using, for instance, a and b. Let R represent a relation. We say that the relation R exists between a and b. Thus:

$$a\,R\,b \qquad (2.2.1)$$

For instance, R might be the relation "greater than." Then (2.2.1) says that "a is greater than b." Other relations we might employ are, for instance, "equal to," "comes before," "is older than," "is a mother to."

### Classification

Relations may be classified as reflexive, symmetric, or transitive.

*reflexive*       A relation is said to be **reflexive** when it applies between the object and itself. Thus:

$$a\,R\,a \qquad (2.2.2)$$

Only a few relations are reflexive. "Equal to" is a reflexive relation; "older than" is certainly not reflexive.

*symmetric*       A relation is **symmetric** when if it applies to two elements a and b in that order, it also applies in the opposite order.

Thus:

$$a \, R \, b \quad \& \quad b \, R \, a \qquad\qquad (2.2.3)$$

Notice that the order in which the elements of the relation are mentioned is of importance in most relations but not in a symmetric relation. "Different from" is a symmetric relation because if two objects are different from each other, it does not matter which we mention first. Notice that "different from" is not reflexive because it isn't possible for something to be different from itself.

Many relations are not symmetric (they are **asymmetric**). "Older than" is a simple example.

*transitive*     First let us state symbolically what a **transitive** relation is:

$$a \, R \, b \quad \& \quad b \, R \, c \Rightarrow a \, R \, c \qquad\qquad (2.2.4)$$

This says that, for a transitive relation, *if* the relation R holds between a and b *and* it also holds between b and c, *then* it must hold between a and c. Thus, if John is older than Harry and Harry is older than Isabelle, then John is older than Isabelle.

It might seem that all relations ought to be transitive. This is not so—"unequal to" is an example. Use this relation and substitute 6 for a, 5 for b, and 6 for c. You can see that "unequal to" is *not* transitive (it is **intransitive**), or we would reach the conclusion that 6 is unequal to 6.

### Visual Presentation

A textual description is necessarily serial, that is, as you read this book, you go from one sentence to the next. It is possible to provide an alternate presentation whereby you need not proceed serially (as in a text with programmed learning). We shall be concerned with many objects and the relations that pertain among them. How can we convey these relations simultaneously? The graph is an answer to this problem.

The graph presents visually both objects and the relations that exist among them with a single picture. The observer can *see* these relations and he is free to examine them in any sequence that he chooses. Hence, graphs provide a way of avoiding the serial (time-wise) presentation inherent in a text.

### Graphs

A **graph** consists of a number of vertices and edges. It must contain at least one vertex. An **edge** is a line segment; a **vertex** is a point. For each edge that appears, the vertex terminating that edge must also be present. Figure 2.2.1 is an example of a graph.

For our purposes, vertices in the graph represent objects and edges represent relations that exist between two of the objects. Since a relation pertains between *two* objects, it is clear why every edge has two terminal vertices in a graph.

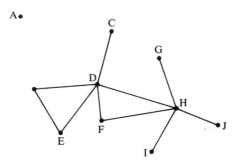

**Figure 2.2.1.** A graph

Of course, mathematicians view graphs as abstract entities. By superimposing meaning upon the vertices and edges, we now have a useful concept for data representation.

### Direction

We have already examined symmetric relations such as "equal to," "next to," and "sibling of." Such relations do not imply a direction; they hold between two elements without regard to which is mentioned first. Symmetric relations are easily represented by the graph as we have described. Then edges represent these symmetric relations and vertices represent the individuals to which the relations apply.

### Asymmetry

A relation is said to asymmetric if it applies between a and b but not between b and a. As examples of asymmetric relations, we have "greater than," "father of," and "uncle of." **Direction** distinguishes which individual is the subject and which individual is the object of the relational verb. Now it makes a difference which individual is mentioned first. We say John is the father of Junior, but the reverse doesn't hold: Junior is not the father of John. The edge does not distinguish which individual is mentioned first.

### Digraph

To indicate which is subject and which is object of the relation, direction is given to an edge, which is then called an **arc**. An arc is shown as an arrow; the subject is at the tail of the arrow and the object is at the head of the arrow. A graph where direction is indicated is called a **directed graph** or, simply, a **digraph**, although arcs still connect the vertices. An example is presented in Fig. 2.2.2.

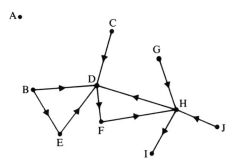

**Figure 2.2.2.** A digraph

### Types of Relations Graphed

Figure 2.2.3 shows the three kinds of relations (reflexive, symmetric, and transitive) conveyed in graph form.

For the reflexive relation, an arrow points from a to a because a relation applies between the element and itself. For the symmetric relation there is an arrow from a to b as well as an arrow from b to a since the relation applies in both directions. For the transitive relation, an arrow points from a to b and another from b to c. The transitivity is indicated by the third arrow from a to c.

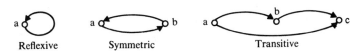

Reflexive        Symmetric               Transitive

**Figure 2.2.3.** Reflexive, symmetric, and transitive

## 2.3 SOME PROPERTIES OF GRAPHS

### Routes

A **route** is a way to get from one vertex to another by passing through edges and vertices of a graph. It is more properly called an

**edge sequence,** a sequence of edges providing a continuous trip from the **initial vertex** to the **terminal vertex.** If our graph represents a file or a list, then an edge sequence represents a search sequence. It is one way to examine records in cells to find a desired record. As you can see, the efficiency of a search depends on

- the file or list structure—its graph;
- the search method—the choice of edge sequence.

In Fig. 2.3.1 we see a graph with many vertices and edges. Suppose we wish to go from a to b. There are many edge sequences we might take. The most direct of these is acdeb. Since there are so many others, we seek some way to classify them. We say that an edge sequence is.

- **elementary** if no edge or vertex appears more than once, e.g., acdeb;
- **simple** if no edge appears more than once, e.g., acdgjkgeb;
- **nonsimple** otherwise.

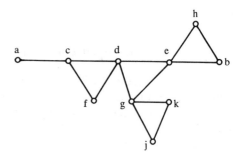

**Figure 2.3.1.** A graph for which we seek an edge sequence from A to B

### Cycle

An edge sequence for which the origin and destination is the same is called a **cycle.** Edge sequences that involve cycles cannot be elementary. For instance, in Fig. 2.3.1, gjkg is a cycle since we leave from g and return to that point.

The example of our simple edge sequence acdgjkgeb contains within it the cycle gjkg. It is simple because no edge is repeated but not elementary because the vertex g appears twice.

Cycles, too, can be classified according to whether they are elementary, simple, or otherwise. The same definitions prevail and, illustrated by Fig. 2.3.2, a **cycle** is

- **elementary** if no vertex or edge appears more than once, e.g., abcda (although the starting and ending vertex is the same);

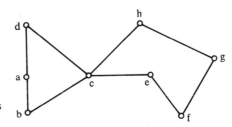

**Figure 2.3.2.** Showing several kinds of cycles

- **simple** if no edge appears more than once, e.g., abcefghcda;
- **nonsimple** otherwise, e.g., abcefghcba.

### Digraph Routes

Many data structures use asymmetric relations and are therefore representable by digraphs. Routes between distinct points of a digraph are called **arc sequences** or **chains** for an obvious reason. The usefulness of graphs for conveying activities in such data structures cannot be underestimated.

Figure 2.3.3 contains a digraph. Again there are many arc sequences between the vertices a and b. In defining such arc sequences, it is important to observe the direction of the arc. We are only permitted to pass along the arc in the direction of the arrow. Thus, in Fig. 2.3.3, acdgeb is an elementary arc sequence from a to b. acdeb is *not* a permissible arc sequence, however, since the arc de does not exist—only the arc ed appears in Fig. 2.3.3.

**Arc sequences** may also be classified as **elementary, simple,** or **nonsimple,** according to whether a vertex or edge is repeated in the sequence.

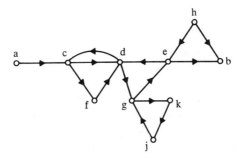

**Figure 2.3.3.** A digraph

### Loops

Loops are the digraph equivalent of the cycle. A **loop** is an arc sequence whose origin and destination is the same. Again, we must keep in mind the directivity of each arc. A loop is defined only when we may pass from the starting point, passing through the various intervening vertices, observing the direction indicated by the arrow on the arc, and thus returning to our starting point. The cycles have been abstracted from Fig. 2.3.3 and are presented in Fig. 2.3.4. Notice that digraphs may have cycles defined after removing the directivity from the graph.

Of the four cycles presented in Fig. 2.3.4 only three are loops. This is so because although there are two ways to leave the vertex h, there is no way to return to h.

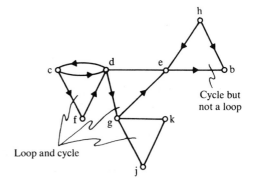

**Figure 2.3.4.** Loops and cycles of Figure 2.3.3

As we would expect, **loops** may be classified as **elementary, simple,** or **non-simple,** according to how many times a vertex or arc is visited.

The graph theorist has defined the loop and this concept is immediately seen to be congruent to "loop" as used by the programmer. Consider the commands in his program as vertices of a graph. Arcs that connect the commands represent sequencing of the computer from one command to the next. A loop in this graph corresponds exactly to a program loop as used in the field.

### Geometry

A graph is a planar drawing but the principles that we apply are not those of geometry. The graph is not viewed as rigid within its plane. Further, the line segments that connect the vertices need not be straight.

Figure 2.3.5 demonstrates this principle. The first three figures are graphically equivalent because each consists of three vertices and three arcs connected in the same sequence. The fact that the arcs are straight lines, circular arcs, or wavy lines has no bearing upon this equivalence. The two circles consisting of three arcs and three vertices at the right of the figure are not equivalent, however, by graph theory! This is so because it is possible to go from a back to a (a loop) for one of them and not the other.

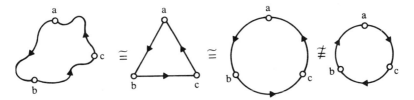

**Figure 2.3.5.** Isomorphic digraphs

## 2.4 TREES

### Importance

A **tree** is a graph that contains no cycles, making it a simpler graph structure. It is a **connected graph** since every vertex terminates at least one edge—there is no vertex from which no edge terminates. A vertex that has no edges incident to it is called an **isolated point** and cannot be found in a tree. Further, a tree with $v$ vertices necessarily contains $v - 1$ edges. An example of a tree is found in Fig. 2.4.1.

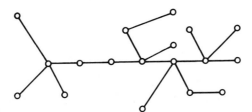

**Figure 2.4.1.** A tree

### Generating

The reason why there are $v - 1$ edges can be demonstrated by seeing how we might go about generating a tree. We start with the simplest tree, a single vertex. We generate more complicated trees by adding vertices. Each time we add a vertex we connect it by an edge to one other vertex. The next step consists of two vertices and the edge connecting them.

Suppose we have generated the tree shown in Fig. 2.4.2. Let us add the vertex h. It is now an isolated point. To become part of the tree we add an edge connecting h to some other vertex. Suppose this is hf in the figure. If we try to add any other edge connecting h to some other vertex, a cycle is created. Thus, if, besides hf, hg is added, we now have the cycle hfgh so that the graph in Fig. 2.4.2 would no longer be a tree.

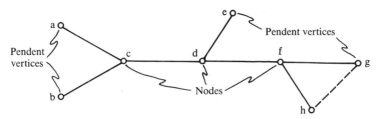

**Figure 2.4.2.** Generating a tree

### Types of Vertices

For a tree, two kinds of vertices are defined by incidence. The **incidence** of a vertex is the number of edges that terminate at this vertex.

- **Pendant vertices** are those that are incident to only one edge—they are hanging.
- **Branch points** or **nodes** are vertices that have at least two edges incident.

In Fig. 2.4.2, a, b, e, and g are pendant vertices. Vertices c, d, and f are nodes. As you can see, they may be considered part of edge sequences connecting other vertices.

### Digraph Trees

Take the structure presented above but consider arcs rather than edges, we have a digraph that is also a tree, a **digraph tree.** This structure is even more important than the tree, for it is a means of representing hierarchical organization.

With regard to the pendant vertices, we now have new terms that describe the direction of the arc with respect to the vertex:

- A **root** is a pendant vertex from which one or more arcs are directed *away* and no arc enters it.
- A **leaf** is a pendant vertex with one or more arcs entering and none leaving.

The branch points or nodes for directed trees are of four kinds, distinguished by whether they resemble a root or a leaf and by incidence. We then have a

- **rootlike node**, which has more arcs leaving than entering;
- **leaflike node**, which has more arcs entering than leaving;
- **linking node**, which has as many arcs entering as leaving;

- **simple linking node**, which has one arc entering and one arc leaving (incidence of 2).

Roots, leaves, and all types of nodes are illustrated in Fig. 2.4.3.

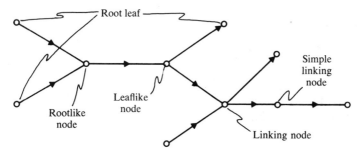

**Figure 2.4.3.** A directed tree

### Arborescence

We have established the directed tree as a simple digraph structure. One that is simpler still is the **arborescence**, a directed tree with a single root (sometimes referred to as a rooted arborescence).

One might consider the root as the *origin* of the directed arborescence. One can find an elementary arc sequence from the root to any leaf. The number of arcs intervening between the root and the leaf is called the **level of the leaf**. A similar definition holds for the **level of a node**—the number of arcs between the root and that node.

The applicability of the arborescence is immediately evident. The organizational chart is such a graph. Here, the root is the president or director. From the root there are one or more arcs to his assistants who occupy the first level. As we go farther away from the root, lower levels of command are found. Finally, the leaves are the workers—no one reports to them. Figure 2.4.4 shows an arborescence.

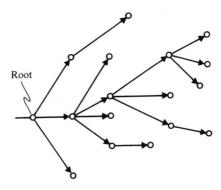

**Figure 2.4.4.** An arborescence

*bifurcating*  A **bifurcating arborescence** contains only the four kinds of vertices that are listed below:

- One *root* is the source of the arborescence.
- The *leaves* are the destinations.
- *Simple nodes* may appear on any level linking the preceding level to the one that follows.
- *Simple rootlike nodes* with only two arcs leaving and one entering (incidence of 3) are the only other kind of node.

Bifurcating arborescences represent simple programs without loops where conditional branches can designate only a single destination. We shall see other applications of bifurcating arborescences. Figure 2.4.5 illustrates a bifurcating arborescence.

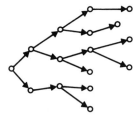

**Figure 2.4.5.** A bifurcating arborescence

### Transitive Graphs

One more simple graph type should be examined. The **transitive** graph is a connected digraph for which cycles may be present but loops are not. This means that transitivity is preserved. Figure 2.4.6 illustrates a rooted transitive digraph.

**Figure 2.4.6.** The transitive graph

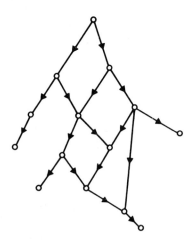

## 2.5 COLORED GRAPHS AS DATA TOOLS

### Colored Graphs

We have examined how a graph can present in pictorial form a number of objects and how each is related to the others in its set. The graph, as it has been described, illustrates how a single relation applies between members of a universe.

Actually, for the same group of individuals, several *different* relations may prevail. Thus, for the family group there are many relations such as "father of," "brother of," "cousin of," and "uncle of" that may prevail among the people concerned. It is interesting that they can all be reduced to two relations—"parent of" and "spouse of"—and all the other relations can be expressed in combinations of these two—birth and marriage. Thus, it is possible to construct a complete family chart if we have some way to convey *both* of these relations.

One way to do this is with a **colored graph** that uses different colors for each of several relations: a red line might indicate marriage and a blue line, birth. Now it is clear how all the individuals in the family are related by observing the color of the arc or the colors in the chain connecting the individuals. We have thus superimposed two graphs that use the same individuals but display different relations.

It is very expensive to use color in a book but, contrariwise, it is essential that two or more relations are presented on the same graph. An expedient is to use lines of different shapes: dashed lines, dotted lines, etc. Thus, our colored graphs are not *actually* colored, but use different kinds of lines.

### Data Objects

The objects of interest to us are generally records and cells. Of course we are also interested in files and lists, but we examine these at the level of record and cell.

To make graphs easier to comprehend, the conventions of Fig. 2.5.1 will be adopted, where records are circles and cells are squares, unless otherwise indicated. An empty cell is distinguished by a square with an "x" in it.

### Data Relations

Figure 2.5.2 shows the various kinds of lines used to convey the relations that might exist between records or cells.

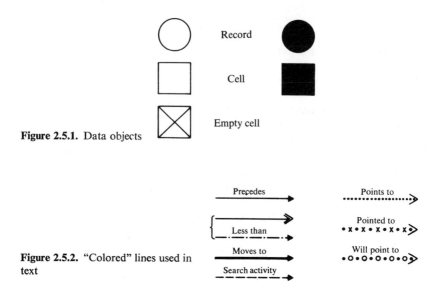

**Figure 2.5.1.** Data objects

Record

Cell

Empty cell

**Figure 2.5.2.** "Colored" lines used in text

Precedes

Less than

Moves to

Search activity

Points to

Pointed to

Will point to

***precede***     The relation of cells to one another is indicated by a light line when the cells are shown as distinct entities, as in Fig. 2.5.3. The cells may be shown by dividing a list, as in Fig. 2.5.4. Then an arc from one cell to its successor indicates the former preceds the latter. Thus in Fig. 2.5.4, since $A_3$ is adjacent to $A_4$ *and* its subscript is smaller, it is clear that $A_3$ precedes $A_4$.

The light line also indicates the current positional relation of records to

**Figure 2.5.3.** Graph of cells in a list

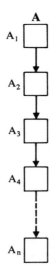

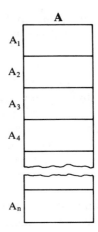

**Figure 2.5.4.** Another graph of cells in a list

one another with reference to the list that holds them. Thus in Fig. 2.5.5 the file a contains records $a_3$, which precedes record $a_4$ in the list.

***less than***     Recall that *order* is determined by the *less than* relation applied to the key field of the records. A record is said to be less than another when the key of the first record is less than the key of the second. A directed thin line from the smaller to the larger is used when no confusion can arise: the dash–dot line is used otherwise.

***points to***     When a record or a cell contains a pointer to another cell (its address), a dotted line is displayed from the pointing record or cell with the arrowhead at the destination cell.

**Figure 2.5.5.** Records in a file

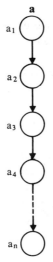

*moves to*    When datum is moved to a location, a heavy arrow has the
              datum at its tail and the location at its head, as shown in
Fig. 2.5.2.

*search*      A search is an examination of a list or file. The search
              proceeds by examining members of the set in some sequence.
a dashed arc indicates the sequence in which the search proceeds. Thus,
although other relations are portrayed statically, a search occurs dynamically
in time and proceeds starting from the tail of the first arrow, examining members of the set in sequence visited by the chain, and ending with the member
at the head of the last arrow.

*changes*     It is possible to indicate alterations being made in a list or
              file on the graph representing it by "coloring" an arc
that is affected by the changes (by using a different kind of line):

- x's are placed on a line to indicate that the relation it represented no
  longer exists.
- o's are placed on the line to indicate a new relation that did not exist
  previously.

We find an example at the bottom of Fig. 2.5.2. The dotted line indicates
a pointer; the line consisting of dots and x's is an old pointer that no longer
exists. The line consisting of dots and o's is a new pointer that didn't exist
previously.

### Data Movement

              A heavy line indicates the movement of data from one place
to another. What is moved and where it is moved is indicated by the shapes
of the vertices.

At the top of Fig. 2.5.6, the contents of cell A are moved to cell B. The
figure indicates that the contents of A are moved because the tail of the arrow

**Figure 2.5.6.** Data movement shown by graphs

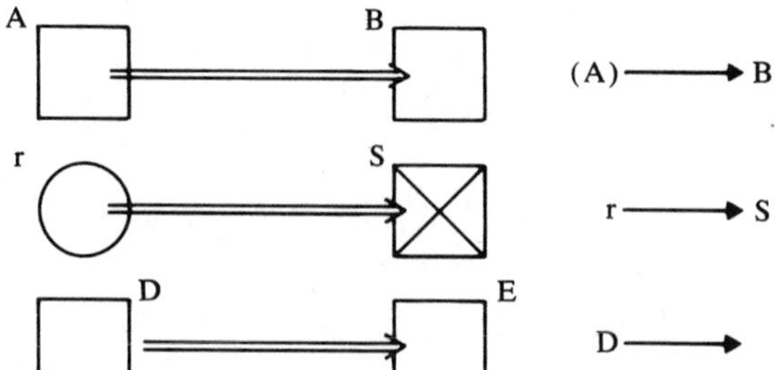

originates within the cell. The figure also tells us that cell B is the destination. The original contents of B will be lost by the transfer.

In the middle of Fig. 2.5.6 we see where the record r is placed in cell S.

At the bottom of Fig. 2.5.6 the *address* of cell D is loaded into cell E. We know that it is the address that is moved rather than the contents of D because the tail of the arrow originates outside the cell boundary.

### Example

We now present an example that shows an important action and also combines some of the graph theoretical concepts that we have learned so far. In Fig. 2.5.7 we see a list, L, consisting of many cells of which only cells A through J are displayed. The x's in cells H through J indicate that they are empty. S is a pointer cell. It contains a pointer to cell H. S keeps track of the remaining empty cells in list L.

The records in list L are in order. This can be discerned by observing the *precede* arrows from A to B, from B to C, etc. There is no arrow from G to H since G is the last record in the file contained in L.

At the left of Fig. 2.5.7 we see a record search t. The purpose of this exercise

**Figure 2.5.7.** Finding where to enter a record in a list

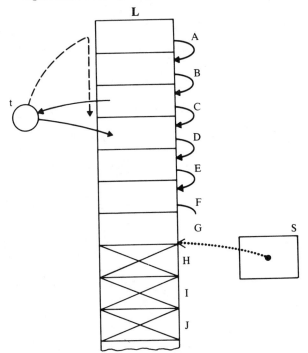

is to insert the record t into the ordered list **L**, so that the resulting list, **L′** is also in order. This requires a search operation of the records contained in **L**. There is a dashed line extending from A to D indicating the direction and duration of the search. The search reveals that the record in C is smaller than t but that t is smaller than the record in D. Thus, t should be placed in the list between the records contained in C and D.

***entry***            We have no place at this time for the record t without reorganizing the list. All records in cells from D to G should be moved down one position; we have then this sequence:

$$(G) \rightarrow H; \quad (F) \rightarrow G; \quad (E) \rightarrow F; \quad (D) \rightarrow E; \quad t \rightarrow D \qquad (2.5.1)$$

Actions up to the last step move records at D through G each down by one cell. Then duplicate records are contained in D and E—since a record at D was moved to E and we still have a copy of it in D—this is the place where *t* will go in the last step.

Notice in (2.5.1) that the sequence of movement is important. The bottom record in the list must be moved to the empty cell first, then the record in the preceding cell, and so forth.

Figure 2.5.8 shows the action required to effect the change in list **L**. Notice

**Figure 2.5.8.** Entering a new record into a list

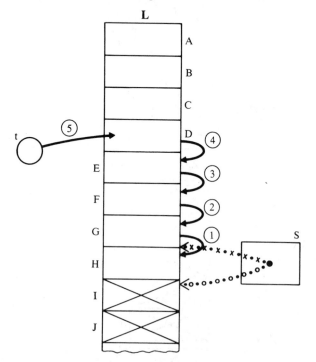

that the heavy lines indicate movement of the records. If only the lines were supplied, we could not tell the sequence in which the movement was done. The small circled numbers attached to the heavy lines show the sequence in which the movement was done:

   ① shows that the record in G was moved to H first.
   ⑤ indicates that t was moved to D last.

S points to the next available free cell: it pointed to H; it now points to I. It is clear how the old and new pointer is conveyed in the figure by different kinds of dotted lines.

**PROBLEMS**

2.1  For *list, cell,* and *element*:
    (a) Explain what each is.
    (b) Indicate what kind of symbol is used for each.
    (c) State what is contained in each.
    (d) Give the symbol that is used for the contents of each.

2.2  A *list, cell,* and *element* have specific addresses in MEMORY. What does this mean? What address is assigned to each? Explain.

2.3  Explain these statements:
    (a) (location) = datum,
    (b) [datum] = location.

2.4  "A *key* uniquely identifies a record." Explain and elaborate on this statement. Why "uniquely"?

2.5  What are the considerations in choosing a key field for a file?

2.6  What does it mean that "one key is larger than another"? How is this determined? How is it represented in symbols? Explain why it may be COMPUTER-dependent.

2.7  How are blanks and numerals in a key handled for System/370? for other computers?

2.8  Order for lists has to do with
    • the relation between cells in the list

                 *and*

    • the relation between records in the file.
    Define carefully for *lists*
    (a) ascending order;
    (b) descending order.

2.9  What is a graph? What do lines and points on the graph represent? How are graphs used?

2.10 How do you conceive a *relation*? What does it mean for a relation to be
(a) reflexive?
(b) symmetric?
(c) transitive?

2.11 For the relations below, state if each is *reflexive*, *symmetric*, and/or *transitive*, and establish with an example or counterexample:
(a) father of,                    (f) higher than,
(b) brother of,                   (g) between,
(c) sibling of,                   (h) smaller than,
(d) equal to,                     (i) precedes,
(e) unequal to,                   (j) loves.

2.12 What is a *directed graph*? What is an *arc*? What does the arrow show? What kinds of relations does the digraph represent?

2.13 What is a route? What is its significance in a list graph? in an organization chart?

2.14 What is the meaning of *elementary*, *simple*, and *nonsimple*? What is a cycle?

2.15 Consider the route from Denver to Chicago. What can you say about it if I go through the same city twice? if I pass along the same stretch of road twice? Which contains a cycle?

2.16 (a) What is the difference between a loop and a cycle?
(b) Can a graph have a cycle and no loop? Explain.
(c) Can a graph have a loop and no cycle? Explain.

2.17 (a) What is a *connected graph*? a *connected digraph*?
(b) Can a digraph be connected and not its graph? Explain.

2.18 In simple terms, what is a *tree*? a *directed tree*?

2.19 For an organization chart, what does a *leaf* and *root* represent? What would more than one root for such a chart mean? Graphically, what is the usual organization chart?

2.20 For the graph of a program:
(a) Must it be an arborescence? Explain.
(b) Must it be bifurcating? Explain.
(c) Must it be transitive? Explain.

2.21 (a) What is a *colored graph*?
(b) Can a digraph be colored? Explain.
(c) Are there colored graphs in this book? How is this possible?

2.22 Explain these data relations, with examples, and classify them (reflexive, etc.):
(a) precede,                      (d) moves to,
(b) less than,                    (e) search,
(c) points to,                    (f) change.

# CHAPTER 3

# Files, Records, and Fields

## 3.1 OCCUPANCY

### What Is It?

Let us suppose that we have a list **L** and that the file **f** is contained in this list. **Occupancy** describes the number relation between the file and the list.

*dense*  The list **L** is said to be **full** or **dense** if each cell of **L** contains a record of **f**. For this to be so, the number of cells in the list must exactly correspond to the number of records in the file. We can state this more succinctly thus:

$$\text{dense:} \quad num\ \mathbf{L} = num\ \mathbf{f} \tag{3.1.1}$$

where *num* is the number operator—it counts the number of records or cells in a list or file.†

*empty*  If no cell in the list contains a record, we say that the list is **empty**. We use a special symbol, $\Lambda$ to indicate "empty," and we say that a list is empty thus:

$$\text{empty:} \quad (\mathbf{L}) = \Lambda \quad \text{or} \quad num\ (\mathbf{L}) = 0 \tag{3.1.2}$$

† Appendix A contains a list of operators used in the text, their definitions, and the place in the text where they are first encountered.

*loose*        A more probable situation is when some of the cells in the list are empty while others contain records of the file. The size of the list is then greater than the size of the file, so that we have

$$\text{loose:}\quad num\ \mathbf{L} > num\ \mathbf{f} \tag{3.1.3}$$

It is important to have some measure indicating the proportion of a list that is occupied. This proportion is called the **occupancy ratio** (OR). This ratio is then the number of valid records in the file divided by the size of the list. In symbolic terms, we have

$$\text{OR} = num\ \mathbf{f}/num\ \mathbf{L} \tag{3.1.4}$$

This ratio can vary between 0 and 1: if OR equals 0, we have an empty list; if OR equals 1, we have a dense or full list.

### Empty Cells

An empty cell is one that does not contain a record of our file. What is the state of the list when we obtain possession of it? Just as when a user takes over a region, that region is full of garbage left over from the previous user, so a list for which we gain possession generally contains somebody else's garbage. Cells that we fill with our records have a definite content known to us. If empty cells have garbage in them, how can we distinguish our record from somebody else's garbage?

To be able to distinguish an empty cell, it must be initialized in some way.

*alternatives*      There are several ways we may initialize a cell—fill a cell before use—so that we can later determine that it is empty:

- enter *blanks* into the cell so that it appears empty;
- enter zeros into the cell;
- set a flag of one or more bits somewhere in the cell;
- enter one or more special characters at the beginning or elsewhere in the cell.

The main problem in distinguishing the empty cell is to make sure that its contents cannot be confused with a valid record. *Blanks* and zeros are generally a good choice, but there are occasions when they won't do. Flags and special characters must appear in places where they wouldn't otherwise appear for a valid record and hence use up valuable room.

*empty*      In what follows we shall use @ to indicate that a cell is *character*   empty, both in the diagrams and in formulas in the text where we might need to test for an empty cell.

Then every list is initialized before we attempt to use it to mark all empty cells with @. Actually @ stands for one of the techniques enumerated above. When a record is removed from a file or is deleted, the cell that it occupies should be marked by @. We find later that we may have to distinguish cells which have remained empty and those which contain a deleted record, and we shall then postulate another symbol (¢) to indicate a deleted record.

### Growth

An application takes place in real time. From one day or month to another, the universe of interest changes—individuals come and go. Our file must reflect this change or growth. At the moment we consider two cases of growth:

- **birth**, where a new record, f′, is placed in some empty cell of our file, $L_j$, indicated thus:

$$f' \rightarrow L_j \qquad (3.1.5)$$

- **death**, where a record is deleted from our file and the cell that it occupies is now indicated as empty. We show deletion of record $f_i$ thus:

$$@ \rightarrow [f_i] \qquad (3.1.6)$$

Notice that a birth record is generated for a new individual who has entered our universe; death indicates that an individual has left our universe.

Births and deaths and the method of handling them depends on the techniques for recording each and the occupancy ratio of our file. When the occupancy ratio is 1, there is really no way to record births. This would cause an explosion of our file unless some alternate method is provided. We shall discuss overflow techniques in Chapter 9 for recording births for dense lists.

## 3.2 ORDER

### Key

Recall that the key identifies a record uniquely. Each record contains a key field which describes one attribute which is unique to each individual. A further requirement of the key field is that an order relation exists among all possible key values. This means, given any two key values, the order relation indicates which of these is smaller.

The order relation may be language- or machine-dependent, it depends on the characters used in the key field and the relation of the codes for these characters.

In some cases a key consists of several fields in the record. But always *we* consider the key as a single field since these multiple fields could be re-arranged and coalesced into a single field.

The key is used for ranking records. We should be able to say that the key of one record is smaller than the key of the second record. Actually, this is somewhat unwieldy to say, and it is simpler to state that a record $a_i$ is smaller than record $a_j$ when we really mean that the key $a_{iK}$ is smaller than the key $a_{jK}$, as in

$$a_i < a_j \Rightarrow a_{iK} < a_{jK} \tag{3.2.1}$$

We do not mean that the length of record $a_i$ is smaller than that of $a_j$.

### Ordered List

We say that a **list is ordered** when a record precedes another only if it is smaller. To put this in symbolic form, we have

$$a_i \prec a_j \equiv a_i < a_j \tag{3.2.2}$$

To restate this, if the key of record $a_i$ is smaller than the key of record $a_j$, then we encounter it first in the file.

$$a_{iK} < a_{jK} \equiv [a_i] < [a_j] \tag{3.2.3}$$

Actually this describes **ascending order**; that is, as we examine the file, we find records of higher and higher key.

For a list, it is the content of each cell that interests us. If cell $A_i$ precedes $A_j$ (because i is smaller than j), then we expect the record found in the first cell to be smaller than the one found in the second cell for ascending order. Thus,

$$i < j \equiv (A_i) < (A_j) \tag{3.2.4}$$

*descending order*    As we pass through the file, if the records that we encounter have smaller keys, then this is described as **descending order.** To write this symbolically, we use (3.2.2) but reverse either the "less than" sign or the "precede" sign, so that we have

$$a_i > a_j \equiv a_i \prec a_j \tag{3.2.5}$$

*loose lists*    A loose list, one with an occupancy ratio less than one, may also be an ordered list. We then mean that if the two cells that we choose are occupied, then (3.2.5) holds. If one or both of the cells that we choose is empty, then we cannot determine from

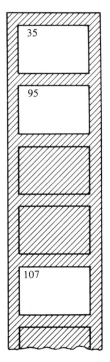

**Figure 3.2.1.** An ordered loose list

them whether the order relation holds. Figure 3.2.1 illustrates this, where we see several records in an ordered loose list. It is possible to write in symbolic form that a loose list is ordered but it is hardly worth the trouble.

One cannot say that an empty list is ordered.

To recapitulate, a list is ordered when it holds the records of the file in sequence, so that as we go from one cell to the next, we encounter records with higher keys or else empty cells.

### Ordered File

How is it possible for a file to exist in a list so that we have an unordered list and yet we can say that the file is ordered? Certainly, as we go from one cell to the next, we do not detect order in the list. There must be some other way of going through the list so that we encounter the records in the proper order.

*pointer*  The means for having an ordered file in an unordered list is an extra field found in each record, a pointer field. The **pointer field** provides a location of the next record in order in the file.

The pointer field is then a **link**—it links records together. The link supercedes the precedence relation of the cells in the list. By following the links, we examine records of the file in the desired order. Hence this is called a **linked list**. Chapter 6 is devoted to the various forms of linked lists and how they are implemented.

### Random List

Records may appear in a list in *no particular order*. Actually, they may be placed there is order of arrival. We might say that there is actually some order in entering the records in the list, but this order depends on outside occurrences rather than the inherent properties of the record. Sometimes this is called a **random list**; it may also be called a **sequential list**. This conveys a subtle aspect of order: as we pass through the list, we encounter records in the same sequence in which they were entered into the list originally. Thus, the sequence we are reconstructing is the original entry sequence of the records into the list.

### 3.3 FIELDS AND FIXED FIELDS

### The Field

The field describes an attribute of an individual. You will recall that a field has both a name and a value. The field name identifies the field within the record; the field value conveys the attribute value of the individual associated with the record.

In this section we describe the field itself—what is it like? There are a number of factors that apply to the field:

- name—the field is uniquely identified by its field name;
- position—generally each field occupies either a fixed position within the record or a fixed relative position with respect to other fields in the record;
- size—the field has a length in bytes or characters that places a limit on the range of values that it may carry;
- variability—the length of a field may vary from one record to the next for a variable length field;
- multiplicity—some attributes may simultaneously have several values and hence a field representing that attribute must have places for several values.

*fixed*        A **fixed field** is one that has a fixed size—it is always the same length. It occupies a fixed relative position and there is no variability in the length since it is fixed. Generally, multiplicity is not associated with a fixed field.

*name*        Fields have names and we shall be using names when discussing alternative formats. A field name hereafter will always appear in capital san serif italics. For instance, a field that conveys the state of residence of the individual might be called *STATE*.

### Attribute Values

Field values somehow represent attribute values. Hence it is important to understand how the attribute value may vary. We may describe variability as

- **continuous**, such as height and weight, which can assume any possible value within a given range;
- **discrete**, such as sex or state of residence, which can only assume a few fixed values.

Regardless of the variability of the attribute, fields can take on only discrete values as prescribed by the field design, due to the limitations of both the data format and the digital nature of the COMPUTER. In the text, field values are designated in san serif capitals. For instance, to indicate that the value of the *STATE* field is New York, we write

$$(STATE) = NY \tag{3.3.1}$$

*range*        The size of a fixed field determines the range of values that field can hold. To design a field we must know the minimum and maximum value that an attribute may have and the granularity of measurement. **Granularity** is the *number* of classes into which attribute values may fall. This is a good way to make continuous data discrete.

This, if we are measuring weight, instead of recording actual weight of an individual, we break up alternative weights into a number of classes, say ten. Then *WEIGHT* will record only ten discrete values and not the actual weight of the individual.

Only certain attributes are amenable to such division. Thus, we cannot properly classify names; the individual's full name must be provided, not a name class. Many seemingly continuous variables, however, can be easily set up in a class system. For example, we have age and income, which lend themselves conveniently to class grouping.

Discrete variables are a natural! Thus we have fifty values for states and two values for sex.

### Field Value Alternatives

The field value represents and portrays the attribute values. The granularity of field value required depends on the application and the attribute being described. Further, the combination used will determine the efficiency with which information is stored.

For instance, consider how to store state information. State names vary in size from, say, four letters (Iowa) to many letters including a *blank* (such as Rhode Island). We shall require a large field to carry all this character information. This could be reduced to two characters and still meet with postal regulations. Thus Alabama is represented as AL and Wyoming is WY. Thus two bytes do the work of many.

There are only fifty states, however, and each can bear a number from 1 to 50. The number itself can be in packed format and will thus require only one byte. Another alternative is to use a single byte of binary information. This byte can take on values from 0 to 255. It is thus useful for attributes that fall into more than 100 classes.

*numeric*     Numeric data often comes into the computer in character
*data*       form where it occupies considerable room, as in the case of
punchcard. When it is being restructured as a record to be stored on an intermediate medium such as tape or disk, however, the data can be compressed. Numeric data can be packed, still keeping its original decimal form so as to occupy half the space of its character form. Further compression can be achieved by converting the information to binary, using either full or half Word format.

### Values of Variable Length

Values of some attributes are represented with a variable number of characters. This may be so because of requirements in the real world or because of an existing system. For example, a full name may require anywhere from six to twenty-six or more characters. This is a requirement of the real world. When an account holder issues a check, the amount of that check may vary from one cent to millions of dollars. We cannot put limitations on the amount of his check except insofar as he should have that amount on hand. This is a requirement of the real world.

A company may use a number of characters to designate each part that it manufactures. This number can vary from one part to another—from, for example, two to eight characters. Besides identifying a part, these characters may also give some of the part characteristics in coded form. This may require twelve or more characters. This is not a real-world requirement but is rather imposed by an existing system.

***in a fixed***    How is it possible to record values of variable length in a
***field***    field of fixed length? After choosing a fixed field of satisfactory size we then have these expedients for putting our
values into the fields:

- **justify**—place the value in the field so that it lines up on either one side or the other;
- **pad**—enter neutral characters in the unused portion of the field;
- **truncate**—eliminate part of the value from the field.

These are displayed in Fig. 3.3.1 and explained below.

### Justify

To justify is to align one side of the value with respect to one side of the field. It is possible to either left justify or right justify and there are good reasons for each.

***alphanumeric***    When we deal with alphanumeric data—data consisting of
***data***    letters, numbers, and symbols—it is customary to left justify,
as demonstrated in Fig. 3.3.1. The reason for this is that it is often necessary to order records or output them alphabetically according to such a field. The "greater than" relation is checked starting at the left for alpha information. If the field were right justified, *blanks* would appear at the left and would give an erroneous comparison between values since a blank is "less than" any other character, which will be "less than" all alphanumerics.

**Figure 3.3.1.** Various values for the NAME field which is of fixed size. Bottom name truncated.

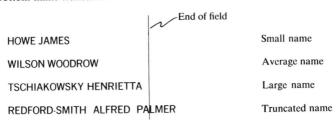

Once an alpha field is left justified, the value may not completely fill the field, as in the upper two entries of Fig. 3.3.1. Again, to realize a proper comparison for ordering, *blanks* are entered into the remainder of the field.

**numeric**    Numbers are right justified. This is true because we often perform arithmetic with these numbers. When two numbers are added into a third field, a carry may be generated. This carry will always take us into the next zero digit to the left. The digit to the right of this position is called the **most significant digit.**

In recording a number that does not completely fill the field after it is right justified, zeros are appended to the left to fill the field. These zeros permit proper calculation and also allow us to compare numbers properly. An example is shown in Fig. 3.3.2.

| | |
|---|---|
| 1234567890 | $\simeq$ $12,345,678.90 |
| 0000666777 | $\simeq$     $6,667.77 |
| 0000000003 | $\simeq$        $.03 |

**Figure 3.3.2.** Numbers are right justified and padded with 0's.

### Truncation

Sometimes an absolute maximum length for a field value is indeterminate or impractical. The name field is the best example. We can allow for a field of twenty-six letters, but you will always run across someone who has an extremely long name. What do we do with him? In the case where we have fixed fields, it is impractical to make the field larger just to accommodate an extremely rare case.

Since it is impossible to fit a large value into a small field, something has to give. One expedient is to discard part of the field value, which is called truncation. We are forcing the value into a field of smaller size, and we discard that portion of the name that interferes least with normal alphabetizing. Of course we shall be unable to distinguish two similar names that differ only in their terminal (rightmost) letters. An example of a truncated name is shown at the bottom of Fig. 3.3.1.

### Field Location

How is a fixed field found? Originally, the designer who lays out the record determines the size of each field. He then assembles these fields into a record and sets the relative position of each field. Since the fields each have a known size, the beginning of each field is a fixed distance from the beginning of the record.

It would be convenient if the record designer placed the fields in the record according to their frequency of use so that later fields are less likely to be referenced. Records often grow like Topsy, however, and nobody likes to take responsibility for just how which fields arrived where.

The programmer, when he writes his application program, would like to reference his fields symbolically. To do this he redefines the record format in his program. In a higher language such as COBOL, he uses the FD or file definition statement; in Assembly language, such as BAL, he defines the record in each field with the DS pseudo. This is described in detail in Section 3.8.

It is possible to name and then symbolically reference each fixed field. Because it *is* fixed, the compiler or assembler will translate each symbolic field into an address in storage so that the proper fields will be manipulated by the COMPUTER.

## 3.4 VARIABLE LENGTH FIELDS

### Need

We have seen that some fields are characteristically fixed in length, such as the Social Security number and state code. Others are characteristically variable in length, such as names, part numbers, and amounts. We have seen how to use fixed fields of maximum value length to hold these variable field values. It should be apparent that, when this is done, valuable record space must be wasted. Only a few actual values fill up all the field length allocated to them. A short eight-letter name, for instance, wastes eighteen characters of a fixed length twenty-six character field.

The way to remedy this is to use fields of variable length. This further complicates matters since necessarily the *records*, too, are now of variable length. Then the program becomes complicated since for *each* field it is necessary to

- find the field;
- determine its length;
- process the variable length field value.

We might think this program will then take longer to run.

Actually, despite the additional apparent complexity, a double saving accrues by using variable length records:

- Space is saved in storing the file on an intermediate medium.
- Time is saved for input and output of the file.

To obtain some idea of this saving, a mailing house that deals with names and addresses of subscribers, which run into the millions, required almost forty reels of tape to store its data in fixed format. When it went to variable record size, it was able to cut down the number of reels by over one-third. Since the time for running the program was consumed mostly by input and output and not by processing, it took a third less time to process the file. Hence a twofold saving was obtained—space and time.

### Techniques

New techniques to handle variable length fields that we shall carefully examine are

- punctuation for delimiting the field;
- a length subfield for describing the length of the value subfield;
- a label that not only delimits the field but indicates which field it is.

*subfield*    The variable length field not only contains a value but a marker or descriptor as well. It is convenient to have a word that indicates these two (or even more) portions of the field. Thus a **subfield** is a reasonable subdivision of a field. Later when we use punctuation to separate fields, we might wish to view the punctuation as a subfield of the field that it demarks.

Another alternative is to maintain the variable length fields as units but to describe them elsewhere. Thus, another separate descriptive field gives the length and/or location of each variable length field.

### Punctuation

One way to distinguish the end of a variable length field is to put a separator after the last character in the field. This separator must be a special character that would not otherwise appear as part of the field value. It need not be a character that prints, but it must be part of the character set. I have called this **punctuation** because the separator is generally some form of punctuation or a special character. To make it more visible, I use $ in examples to indicate the end of a field. Names do not usually include $; it is rare to find even a part number that includes $. Therefore, the occurrence of $ can properly indicate the end of a field or subfield without being confused with the field value.

It is usually a requirement that the *relative* position of the fields in the record be maintained. Thus, as we sequence from one variable length field to another, there need be nothing inherent in the field to identify it; it is identified by relative position.

**Figure 3.4.1.** Variable and fixed fields separated by $

In Fig. 3.4.1 we see a number of variable and fixed length fields combined in a record. The end of each field is signaled by $. Notice that *NAME* contains an embedded *blank*; this does not indicate the end of the field but is a proper part of it.

When fixed and variable length fields are intermixed, $ is required to indicate the end of a variable length field, however, the length of a fixed length field is (or should be) known to the programmer. He can sequence through the fixed number of characters that comprise this field and know that, as he leaves this field, a new field follows, starting with the next character. Hence, he would omit $ for the fixed fields, as shown in Fig. 3.4.2.

**Figure 3.4.2.** Only variable fields need be terminated by $.

```
NAME  NUMBER  HOURS  SEX  DEPT  CLASS
JIM HOWE$03729384396$M0342132
```

### Length Subfield

A **length subfield** may be provided to indicate the length of the value subfield that follows. Thus, each variable length field contains two subfields:

- The *LENGTH* subfield specifies the length of the datum.
- The *VALUE* subfield contains the value of the datum.

For this system to work, the following requirements must be met:

- The *relative* position of each variable length field within the record is fixed.
- Those fields that have variable length are known to the programmer as such.
- The *LENGTH* subfield is of fixed size.

Figure 3.4.3 presents an example using the *LENGTH* subfield. Notice *NAME* has two subfields. *NAME LENGTH* is a fixed subfield one byte long. In the figure, this subfield contains the value 8. This indicates that the field called *NAME VALUE* is eight bytes long. The reader can follow a similar example for *HOURS*, containing *HOURS LENGTH* and *HOURS VALUE*.

When fixed length fields intervene, their length is known, and so no length descriptor is associated with them.

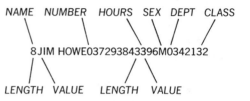

**Figure 3.4.3.** The LENGTH subfield records the *length* of the subsequent variable length VALUE subfield.

### Separate Length Field

Here again a length descriptor is provided for each variable length field; however, this length is not physically incorporated into the field it is describing. Instead, all these length descriptors are gathered together into a single field. This might be called the *LENGTH* field. It contains one fixed length subfield for each variable length field in the record.

In Fig. 3.4.4 we see that field *LENGTH* contains the subfields *NAME LENGTH, HOURS LENGTH,* etc. The fields to which they apply are not subdivided but contain only the field value. Thus *NAME* is an eight-byte field containing only the name itself.

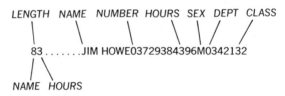

**Figure 3.4.4.** Field length may be kept in a separate LENGTH field with subfields such as NAME LENGTH and HOURS LENGTH for each variable length field in the record.

It is possible to mix fixed and variable length fields. No length descriptor is associated with the fixed length field since the programmer knows its length.

*pointer*    Although it takes up more space to provide a pointer to a variable length field, it makes it easier to find that field. Thus, a *POINTER* field might provide a *LOCATION* and *LENGTH* subfield for each variable length field.

### Inclusive Versus Exclusive Length Descriptors

As we have postulated a length subfield, it gives the length of the value subfield that either follows or is at some remote point. As presented, the length provided does not include the length of the length subfield itself. In some cases, it is advantageous to make the value in the length sub-field **self-inclusive**. For instance, if the length of the subfield called *LENGTH* is one byte and the length of the *VALUE* subfield is eight, then the value in *NAME LENGTH* would be nine instead of eight. This has an advantage in that we can accumulate lengths as we sequence through a record. To find the beginning of the *next* field, simply add the contents of the length subfield of *this* field to the address of *this* field.

### Labels

If each field is preceded by a label, then the field is not only marked but identified. This means that the order in which the fields appear in the record is no longer required to be relatively fixed since each bears an internal label. The problem is that if we use conventional alpha characters for the labels, then *every* field must begin with a label or else it would become confused with the characters that comprise the field value.

Since eight-bit codes provide 256 different combinations, many of these are available as field labels. The fact that they are not printable characters is a helpful by-product.

Figure 3.4.5 shows an example where the fields of our record are each pre-fixed with a nonprinting label character. This accounts for the circle around the character so as not to confuse it with a letter. The final field in the record must have another special character to indicate its termination; $ is used here for that purpose.

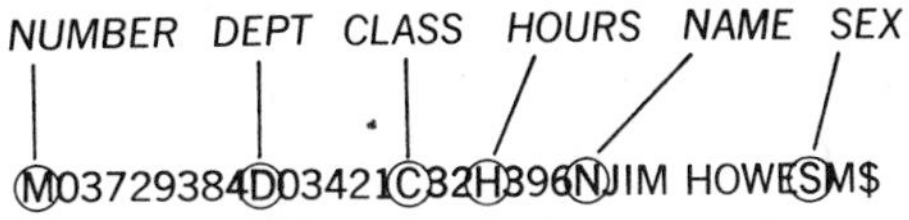

**Figure 3.4.5.** Using labels to delimit fields where Ⓝ, Ⓜ, etc. are special, nonprinting symbols

NUMBER DEPT CLASS HOURS NAME SEX

Ⓜ03729384Ⓓ03421Ⓒ32Ⓗ896ⓃJIM HOWEⓈM$

## 3.5 MULTIFIELDS

### Need

Individuals have some attributes which are repetitious and which apply in varying number according to the individual. Some examples, prefixed by application type, should clarify this:

- personnel—when requested to list previous residences over the past ten years, the applicant may have none, one, or many listed;
- bank account—an individual's record for the month indicates with-drawals and deposits; either of these may be lacking or may be present in varying numbers;

- retrieval—associated with each book in the library is at least one but probably several descriptive words under which it is cross-indexed in the topic file;
- personnel—an employee is usually required to supply name and relation of each dependent for which a deduction is being taken.

As you can see, for each of these attributes it is desirable to have room in the corresponding field to hold several distinct field values. A single name describes the field, e.g., residence, activity, topic, dependent. I prefer to call this collection of field values simply a field and to consider the space occupied by each value a subfield. Notice that the number of subfields can vary between zero and some, possibly undefined, maximum value. In Section 3.6 we examine the case where a value for the field is entirely absent; here let us simply examine the case of a field with one or more subfields, which we now define.

### Definition

A **multifield** is a field that consists of multiple value subfields, each representing a different value for an attribute that has a single attribute (and field) name.

### Techniques

For this to be a useful concept it must be possible to distinguish one value subfield from another and to know which is the last subfield of the multifield. Different techniques prevail according to whether the subfields are of variable or fixed length. As with fields, fixed length subfields are easier to deal with but are less economical in terms of space. Thus we pay for the variable length subfield in the extra technique required to determine its length and position, gaining economy of storage in return.

We have available the previously described techniques:

- a number descriptor to indicate how many subfields comprise the field;
- separators such as punctuation to delimit subfields;
- length subfields associated with each value subfield.

### Number

To describe the fixed multifield where the length of each subfield is the same, we need only know the number of subfields that the multifield contains. Now, even though we have fixed length subfields, the length of the field is variable and depends on the number of subfields contained in it.

The *NUMBER* subfield is fixed in length and hence has a maximum value determined by the record designer and usually a nibble or, at most, a byte is sufficient. The first alternative places the *NUMBER* subfield first in the field, followed by the indicated number of *VALUE* subfields. This is illustrated in Fig. 3.5.1. The value for the preceding fixed *BRANCH* field is J. The *AMOUNT* field begins with a *NUMBER AMOUNT* subfield, the value of which is 3. This indicates that three values follow, as are shown. The fixed field called *DATE* is next.

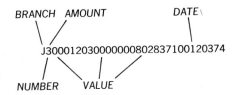

BRANCH AMOUNT                    DATE

J3000120300000000802837100120374

**Figure 3.5.1.** Fixed multifield with number subfield

NUMBER    VALUE

A variation of this technique is to provide a separate *NUMBER* field that contains one subfield for each multifield in the record. This is illustrated in Fig. 3.5.2. Here, at some remote location in the record, we find *NUMBER*, the first subfield of which is *AMOUNT NUMBER*, the value for which is 3. *AMOUNT* follows immediately after *BRANCH* and begins with the first value subfield since the number of values is found elsewhere.

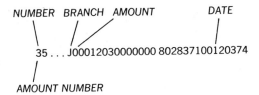

NUMBER  BRANCH  AMOUNT          DATE

35 . . . J00012030000000 802837100120374

**Figure 3.5.2.** NUMBER field contains one subfield for each multifield in the record.

AMOUNT NUMBER

### Punctuation

For variable length subfields, we can use punctuation for indicating the end of each subfield. One technique uses a different separator for subfields than for fields.

Figure 3.5.3 uses % to separate subfields and $ to separate fields. Now it is not necessary to convey the number of subfields since when $ is reached, processing terminates. In the figure, *AMOUNT* begins right after *BRANCH*. After each subfield we find %. Following the last %, we find $, after which comes *DATE*.

BRANCH AMOUNT
DATE
J12030%7%2837100%$120374

**Figure 3.5.3.** A subfield symbol % separates variable length subfields in a multifield.

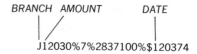

If, as we examine a subfield, we come to a field terminator ($) instead of a subfield terminator (%), it is clear that the subfield also terminates. Hence in Fig. 3.5.4, % associated with the third subfield is omitted.

In some cases it may be useful to have the same separator for both subfields and fields. Then how can we determine that the field, not just the subfield, is ended? One answer is presented in Fig. 3.5.5. Here, $ separates each subfield, but $$ terminates the last subfield and indicates the end of the field. Simple variable length fields can still use a single $ for terminators.

BRANCH AMOUNT
DATE
J12030%7%2837100$120374

**Figure 3.5.4.** The terminal subfield separator may be dropped in favor of the field separator.

BRANCH AMOUNT
DATE
J12030$7$2837100$$120374

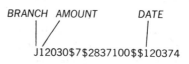

**Figure 3.5.5.** A single separator $ when used for subfields can be doubled for multifield terminator.

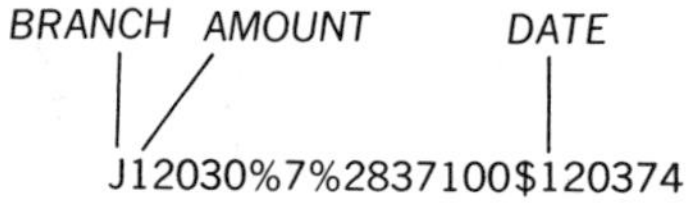

## Length

We have seen that another way to describe variable length fields is to associate a length subfield with each. Why not do the same for variable length subfields—provide a length subfield for each variable length subfield? Also, for a multifield, as for other fields, we are concerned with where the field ends. Again, this may be determined by another length subfield. Then each such multifield has one length subfield for the field proper and one length subfield for each value in the field.

In the example of Fig. 3.5.6 we find *AMOUNT* after *BRANCH*. *AMOUNT* begins with *FIELD LENGTH*, indicating that the number of bytes in the field is sixteen. Next we find a *SUBFIELD LENGTH* of five, which applies to the following five-byte *VALUE* subfield. Another *SUBFIELD LENGTH* follows, containing one, which applies to the succeeding *AMOUNT VALUE*

BRANCH AMOUNT
DATE
J1651203017728371001 20374
FIELD LENGTH
SUBFIELD LENGTH

**Figure 3.5.6.** A field length begins this field and each value subfield is prefixed by its own length.

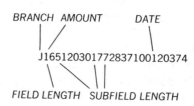

subfield. The last *SUBFIELD LENGTH* follows, which contains seven, and applies to the last *AMOUNT VALUE* subfield that follows. Since the program has been counting bytes and they are all used up, it knows that the next field *DATE* follows.

### Number and Length

Since *NUMBER* counts subfields contained in a field, it is always smaller than the length of the field. Another alternative is to indicate how many subfields are present and then associate a *SUBFIELD LENGTH* with each subsequent value subfield. This alternative is demonstrated in Fig. 3.5.7, which shows how self-inclusive lengths are used.

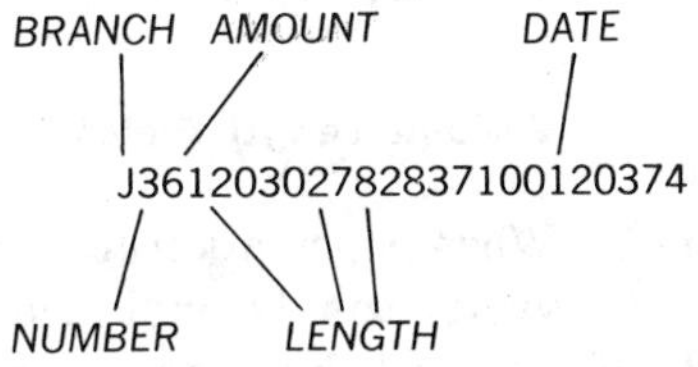

**Figure 3.5.7.** A number subfield followed by three self-inclusive length subfields completely describes this multifield.

*AMOUNT* begins with *AMOUNT NUMBER*, which is 3, to convey that three subfields follow. Next we find the *SUBFIELD LENGTH*, 6, which indicates that there are six bytes including this one for this subfield, etc.

There are other alternatives where *SUBFIELD LENGTH* is disassociated from the *SUBFIELD VALUE* to which it applies and is listed separately. Still more alternatives should be apparent to the reader.

## 3.6 ABSENT FIELDS

It may not be immediately clear which fields might be absent. Fields represent attributes—how can an attribute be "missing"? Certainly, this is true for most attributes. How could we expect our employee to lack a name, a height, or a weight?

It is appropriate for some attributes to be missing. Generally missing attributes are not physical (although there are cases when even this is true—consider hair color for the bald person). Usually the attribute is operational—whatever that means. For instance, our employee may have *no* dependents. He may also have *no* previous residences within the past ten years (or ever). For our savings account, we may have made *no* withdrawals this month.

When the record contains fields for which the attributes may be absent, the record may convey this by providing a null value for the field or in some other way as described below.

### Fixed Field

A fixed-size record cannot tolerate a missing field that would hence reduce the record size. For fixed records using fixed fields, therefore, we must indicate within the field that the attribute the field represents is absent.

When the attribute is missing, the fixed field is set to some unique value. This may be zero if we would not expect such a value in this field. Otherwise, some unique combination is used. Further, to make this a useful concept, the program must check the field value every time it examines this field to determine if the attribute is missing by detecting this "absent" combination.

### Variable Length Fields

***punctuation***     When punctuation delimits subfields and fields, missing field values can be indicated by supplying the terminal field marker only. Figure 3.6.1 shows an example where % separates subfields and $ separates fields. Following *BRANCH*, we find *AMOUNT*, which contains only $, indicating that there is nothing else in the field. *DATE* immediately follows $.

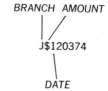

BRANCH  AMOUNT

J$120374

DATE

**Figure 3.6.1.** *AMOUNT* uses % and $ as subfield and field separators but is absent.

When a variable length field is delimited by a length specifier, a length subfield set to 0 indicates that the value subfield is absent. It is difficult to find an example for a variable length field that is not a multifield and might also be absent, but this technique is illustrated in the following subsection.

### Multifields

***punctuation***     When punctuation is used to separate both subfields and fields, the presence of only a terminal field symbol indicates the absence of the field. An example where % separates subfields and $ separates fields is presented as Fig. 3.6.1.

We have seen the use of a single symbol ($) to separate subfields and the same symbol appearing doubly ($$) to indicate the end of a field. Now, to convey the absence of such a field, only the terminal double symbol appears, as in Fig. 3.6.2.

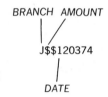

BRANCH   AMOUNT

J$$120374

DATE

**Figure 3.6.2.** *AMOUNT* uses $ and $$ as subfield and field separator.

*count*    We have seen how a multifield can be specified with a count and auxiliary lengths. Such a field is signified as absent by setting the count to zero. Thus, in Fig. 3.6.3, *AMOUNT NUMBER* contains zero and that value both starts and terminates the field; the next character belongs to *DATE*.

BRANCH   AMOUNT

J0120374

NUMBER   DATE

**Figure 3.6.3.** When *AMOUNT NUM-BER* is 0 it shows that *AMOUNT VALUE* is absent.

The same technique is used when the count is established as a subfield of a field that contains counts for several multifields. In Fig. 3.6.4, *NUMBER* contains the subfield *AMOUNT NUMBER*, which is 0. In this case *AMOUNT VALUE* is missing and *DATE* comes right after *BRANCH*.

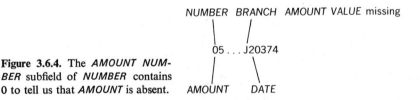

NUMBER   BRANCH   AMOUNT   VALUE missing

05 . . . J20374

AMOUNT       DATE

**Figure 3.6.4.** The *AMOUNT NUM-BER* subfield of *NUMBER* contains 0 to tell us that *AMOUNT* is absent.

*length*    If the length subfield is set to indicate a field length of 0, this indicates that the value subfield is absent. Figure 3.6.5 presents an example using exclusive lengths. *AMOUNT LENGTH* of 0 indicates that nothing follows for this field and *DATE* occurs next.

When *LENGTH* is a self-inclusive descriptor, then its value appears as

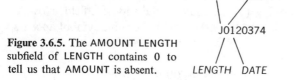

**Figure 3.6.5.** The AMOUNT LENGTH
subfield of LENGTH contains 0 to
tell us that AMOUNT is absent.

nonzero in the record. The value that appears there is simply the length of
the length subfield itself.

## 3.7 SUBRECORDS

### Need

Sometimes a set of fields is associated together and none,
one, or more of such sets of fields may appear in the record. We could name
this collection of fields, calling it perhaps a superfield. Another way to consider it is like a miniature record. This is my preference, and such a collection
is called here a **subrecord.**

It is hard to obtain an adequate feeling for a subrecord in terms of physical
quantities. If it were possible to divide the individual into subindividuals,
then one subrecord would correspond to each subindividual. It is better to
consider the subrecord with reference to a construct for the individual. Thus,
part of the description of a person is his dependants. For each dependant
we might need to know the name, the relation to the individual, the age, and
the residence. All this information about the individual's relative comprises one subrecord. The individual is really a constellation that includes his
relatives and their description.

**Figure 3.7.1.** Part of an invoice

| Quantity | Item no. | Description | B/O | Price | Extension |
|----------|----------|-------------|-----|-------|-----------|

**Example**

We shall use the invoice as an example of a record containing subrecords. A portion of an invoice is presented in Fig. 3.7.1. The invoice contains much information about the customer, the salesman, and so forth. Of most interest to us are the items ordered by the customer. A number of fields are required to describe each item and its disposition. Thus in the main portion of the invoice we see the following information for each item ordered:

- quantity—the number of pieces ordered;
- part number—the abbreviated identification;
- description—a verbal description;
- B/O—to indicate whether the item is shipped or back ordered;
- price—the unit price;
- extension—the price for the number of pieces ordered.

Each of these is allocated a separate field of the subrecord.

**Alternatives**

The record designer should know which alternatives are available for his subrecords. Actually, each field we have examined above is a candidate for a field in a subrecord. We can have fields of variable or fixed length. We may use multifields. Fields may be absent.

Of course the usefulness of the subrecord is that several such sets of fields may appear in the record. It is also conceivable that a subrecord may be absent. Consider the employee with no dependents!

**Subrecord Format**

Figure 3.7.2 shows a typical subrecord format. We should think of the subrecords as a group with some cohesive characteristic. Then associated with this group is a number—the number of subrecords provided. Thus in the figure, we see that the subrecord group begins after the field *CUSTNO*. The first field we encounter is *SUBRECNO*, which conveys the number of subrecords in the group, in this case three. In this case *SUBRECNO* is a fixed field and no delimitor appears for it.

The first field of the first subrecord in *QUANT*, corresponding to the quantity ordered. The subrecord consists of a fixed number of fields, some variable

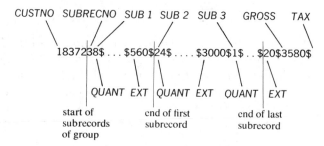

**Figure 3.7.2.** Multiple subrecords in a record

in length, some of fixed length, but each terminating with $. Thus the program to sequence through the subrecord can be fairly uniform. The end of one subrecord and the beginning of the next is determined by the program by counting fields.

After the last subrecord of the group, other fields may follow. For instance, in the figure we see that the gross cost *GROSS* follows the last subrecord. This is followed by *TAX*, etc.

Figure 3.7.3 shows the inside of a subrecord. The reader can discern the fixed and variable length fields. Notice that *B/O*, the back order field, can be

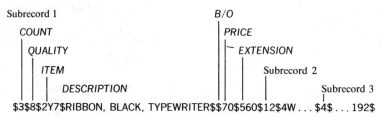

**Figure 3.7.3.** A subrecord with fixed and variable length subfields, some of which may be absent using punctuation

absent to indicate that the ordered quantity of this item was actually shipped. B/O will be present only if some quantity item is back ordered.

### Relation to the Original Data

At the top of Fig. 3.7.4 a typical invoice is shown; at the bottom, how this information would be reorganized to be entered into a variable-sized record.

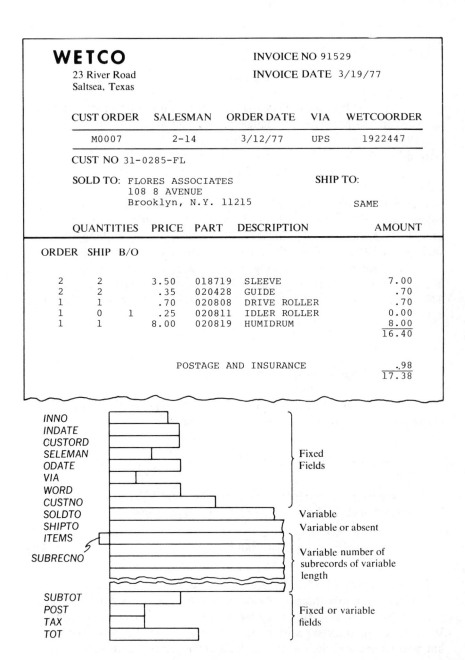

**WETCO**
23 River Road
Saltsea, Texas

INVOICE NO 91529

INVOICE DATE 3/19/77

| CUST ORDER | SALESMAN | ORDER DATE | VIA | WETCOORDER |
|------------|----------|------------|-----|------------|
| M0007 | 2-14 | 3/12/77 | UPS | 1922447 |

CUST NO 31-0285-FL

SOLD TO: FLORES ASSOCIATES      SHIP TO:
108 8 AVENUE
Brooklyn, N.Y. 11215      SAME

| QUANTITIES | | | PRICE | PART | DESCRIPTION | AMOUNT |
|---|---|---|---|---|---|---|
| ORDER | SHIP | B/O | | | | |
| 2 | 2 | | 3.50 | 018719 | SLEEVE | 7.00 |
| 2 | 2 | | .35 | 020428 | GUIDE | .70 |
| 1 | 1 | | .70 | 020808 | DRIVE ROLLER | .70 |
| 1 | 0 | 1 | .25 | 020811 | IDLER ROLLER | 0.00 |
| 1 | 1 | | 8.00 | 020819 | HUMIDRUM | 8.00 |

16.40

POSTAGE AND INSURANCE      .98
17.38

INNO
INDATE
CUSTORD
SELEMAN
ODATE     Fixed
VIA     Fields
WORD
CUSTNO
SOLDTO     Variable
SHIPTO     Variable or absent
ITEMS
    Variable number of
SUBRECNO     subrecords of variable
    length

SUBTOT
POST     Fixed or variable
TAX     fields
TOT

**Figure 3.7.4.** The relation of the invoice to the subrecord

## 3.8 RECORD STRUCTURE

### Records and Language

It is not immediately apparent that the structure of a record depends on the programming language used for record definition. Unfortunately, this may be true. Many higher level languages or language dialects only allow for fixed fields. If all the fields are fixed in length, so is the record. We must then resort to padding and truncation to hold variable field values.

This chapter has indicated the savings that can accrue by variable length records using variable length fields. Certainly this should be attractive to the user, but it somehow loses its appeal if common programming languages such as COBOL cannot accommodate variable length fields. Fortunately there is no problem when assembly languages are used. Examples that appear in the remainder of this section use BAL, the assembly language for System/370. The language is not explained in this text, but the use is not difficult to figure out.

### Fixed Records

For fixed length records, we can state symbolically with the length operator *len* that all records have the same size:

$$len \; r_i \; = len \; r_j \qquad \text{(all i, j)} \tag{3.8.1}$$

so that "the length of records $r_i$ and $r_j$ are the same." This implies that similarly situated fields in different records have the same size:

$$len \; r_{im} = len \; r_{jm} \qquad \text{(all i, j, m)} \tag{3.8.2}$$

This makes it very easy for the programmer to handle.

Generally, the first thing a programmer does is set up a work area in his program for the record. He knows the type, location, and length of each field. When he sets up the work area, he not only names it for the record concerned but also names and defines each field. For instance, a card image may be named and defined as below.

```
REC      DS 0CL80      DEFINE RECORD
KEY      DS PL4        PACKED KEY, 4 BYTES
NAME     DS CL26       NAME + BLANKS, 26 BYTES
HOURS    DS CL3        . . .
              .
              .
              .
CLASS    DS CL2            . . .                    (3.8.3)
```

Now to reference any field in the record he uses its symbolic name. Thus with the two commands below, he *moves* RATE to a work area called WORK and then *multiplies* it by HOURS.

```
MVC    WORK,RATE        MOVE RATE TO WORK
MP     WORK,HOURS       MULTIPLY BY HOURS          (3.8.4)
```

### Record Length

For fixed records, the record length is known. For variable length records, the record lengths differ from one another. For the programmer to handle a variable length record he must know its length. He requests a record from the Access Method, using some imperative such as GET. He receives either the record or its location. The AM in turn received the record in a buffer. He too had to know how large the record is so that he could provide space for it. Generally the means for doing this is a record length designator that precedes the record, as shown in Fig. 3.8.1. This enables both the user and the Access Method to provide space for the record.

### Field Positions

We have seen that variable and fixed length fields can be intermixed within a single record. Where this is done, we do not have the advantages of the fixed field. Recall that fixed fields also have fixed locations within the record and hence can be addressed symbolically. Why not put all the fixed fields first and thus accrue this advantage? Then, any fixed field is handled in the variable record just as though it were in a fixed record. Also, why not use a single method for recording variable length fields so that the technique for handling a variable length field is always the same? This being

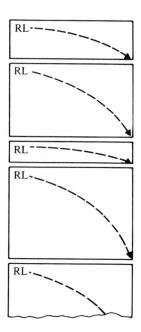

**Figure 3.8.1.** The record length, RL, describes the length of the variable record.

the case, let us now see how to handle different kinds of variable length fields, assuming that the first of these starts at a fixed position in the record.

### Using a Length Field

Let us suppose that we have a one-length field. Call it **LEN**, and suppose that it contains length subfields for each variable length field. To start, load the address of the length field into a GPR, say GPR7,

$$\text{LA} \quad 7,\text{LEN} \tag{3.8.5}$$

This enables us to obtain the length of the first variable length field and put it into GPR8:

$$\text{L} \quad 8,0(7) \tag{3.8.6}$$

We have housekeeping to do so that GPR7 now records the position of the next length subfield with

$$\text{A} \quad 7,=\text{F'1'} \tag{3.8.7}$$

or

$$\text{LA} \quad 7,1(7) \tag{3.8.8}$$

Now suppose that the starting position of the variable length field is known and has been put into GPR9. We can move this field to a work area by setting up a command somewhere in the program called MOVE1, which looks like this:

$$\text{MOVE1} \quad \text{MVC} \quad \text{WORK(0),0(9)} \qquad (3.8.9)$$

Notice that the length of the field to be moved is specified as 0; that is because we don't know its length at the time of writing. The execute command with mnemonic EX will cause the execution of this command and will also enter the length from GPR8 if we specify it thus:

$$\text{EX} \quad \text{8,MOVE1} \qquad (3.8.10)$$

When we are ready to examine the next variable length field, we can easily find its beginning. It appears right after the field we have been working with, the address of which is in GPR9, and the length of which is in GPR8. Simply add these two quantities together, thus:

$$\text{AR} \quad \text{9,8} \qquad (3.8.11)$$

### Using the Length Subfield

As before, suppose that we have picked up the start of the variable length field in GPR9. We load the length subfield, which we assume is nonself-inclusive, placing it in GPR8, thus:

$$\text{L} \quad \text{8,0(9)} \qquad (3.8.12)$$

To find the start of the *VALUE* subfield, we add the length of the fixed length subfield *LENGTH*. Assume that it is one byte long. The location that we desire is placed in GPR9, thus:

$$\text{LA} \quad \text{9,1(9)} \qquad (3.8.13)$$

Assume that a MVC command exists at MOVE1 as in (3.8.9). We request that it be executed thus:

$$\text{EX} \quad \text{8,MOVE1} \qquad (3.8.14)$$

The start of the next variable length field is simply calculated with

$$\text{AR} \quad \text{9,8} \qquad (3.8.15)$$

### Using Punctuation

When our record distinguishes fields or subfields by means of punctuation, it is necessary to search for and find the desired special symbol. Luckily, in BAL, Assembly Language for System/370, there is a *translate and test* command TRT that does this search operation for us. We provide two operands and give their location in the command. The first is a table that contains all the characters in our alphabet. For each character that might appear in a field, the table contains 0's; for the punctuation used as delimiters, the table should contain a special character that identifies the particular kind of punctuation encountered. The second operand is the place in the record where search should begin.

To use TRT, let us suppose that our present position in the record is contained in GPR9 and that the symbolic name for the location where the table is stored is TBL. The command is issued thus:

$$\text{TRT} \quad \text{TBL,0(9)} \tag{3.8.16}$$

During execution of the command, the COMPUTER checks each character against the entry in the table. As long as the COMPUTER finds 0 in the table, it goes on to examine the next character. When the character encountered has a nonzero entry in the table, the address of that character is placed in GPR1 and the entry in the table is placed in GPR2.

For our purpose, we execute the command at (3.8.16) and when a punctuation symbol is found, this location is placed in GPR1 and an identifier in GPR2. This identifier may be checked but we omit this step.

We now have the absolute address of $ in GPR1; it should be in GPR8. We put it there with

$$\text{LR} \quad \text{8,1} \tag{3.8.17}$$

The length of the field is the difference between the address of $ and that contained in GPR9. Thus, subtraction is performed with

$$\text{SR} \quad \text{8,9} \tag{3.8.18}$$

We can now execute the move with

$$\text{EX} \quad \text{8,MOVE1} \tag{3.8.19}$$

It is easy to see how the start of the next field is found.

### Subrecords

Subrecords will be of varying length and they should be found in the second half of the record, which contains the variable length fields. Each group of subrecords starts with a group number that is extracted by the program and is then used to indicate how many subrecords to process through. Thus subrecord processing can be considered like variable field length record processing in miniature. If possible it is better to place the fixed fields first in the subrecord, but generally this goes against the intent of the application and mixed fields are expected. It is possible to use punctuation to terminate even fixed fields and thus make processing more uniform.

## PROBLEMS

3.1 Explain the terms *dense*, *loose*, and *empty* with respect to a list. How are they applied to a file? Explain.

3.2 (a) What is the *occupancy ratio*? Define symbolically.
    (b) What is a *hole*? What is its significance?

3.3 Explain different ways to identify *holes* in a list using the binary representation for System/370 (or your installation computer):
    (a) a key of 0's,
    (b) a key of blanks,
    (c) hex FF starting the key,
    (d) a tag—which? where?

3.4 Distinguish and define *birth*, *death*, *change*.

3.5 (a) How does an ordered file differ from an ordered list?
    (b) Explain how to keep an ordered file as a loose list.
    (c) How can an ordered file be in an unordered list?

3.6 What is a *sequential list*? Is it ordered?

3.7 For a fixed field:
    (a) What is *fixed* about it?
    (b) When will it start at a fixed absolute position in the record?
    (c) When will it start at a variable absolute position from one record to the next?

3.8 (a) Explain how a continuous variable can be represented by a fixed field.
    (b) Construct a one-byte height-in-inches field, and explain a "natural" allocation of field values for packed (two digits per byte) and binary (eight bits).

(c) Assume that average heights vary between 55 and 85 in. How can field values be assigned to attribute values to achieve the best *granularity* (distance between adjacent recorded values) for
(i) packed,
(ii) binary.

3.9   Explain how variable length data can be stored in a fixed field.

3.10 (a) Why left justify alphanumeric data?
(b) Why right justify numeric data?

3.11 (a) Why pad alphanumeric data with blanks?
(b) Why pad numeric data with zeros?

3.12 What is truncation? What is its disadvantage? What is the main source of objection to it?

3.13 (a) How does the programmer refer to a fixed field in a fixed record? Explain why it is so easy.
(b) How does the programmer locate a fixed field in a variable record? Why can't the programmer simply use its name?

3.14 (a) Explain the advantage of the variable length field.
(b) What effect does the variable length field have upon the record containing it?

3.15 (a) What is a subfield, and why is it so-called?
(b) If punctuation is considered a subfield, must a variable length field always contain subfields?
(c) When? Explain.

3.16 The following names are to be placed into a key field so that the records they identify can be arranged alphabetically:

JIM LO
ALFRED HENRY SMITH
ALFRED HENRY SMITH JR.
MARKANTONIO GILBERTO VILLALOBOS
THE AMERICAN SOCIETY FOR THE PREVENTION OF CRUELTY TO CHILDREN
Show how each is represented by
(a) fixed field of twenty-six letters left justified, padded with blanks and truncated;
(b) variable field with $ for separator;
(c) self-inclusive length prefix.

3.17 Why must the length subfield be of fixed length?

3.18 Contrast the self-inclusive and nonself-inclusive length subfield for
(a) ease in sequencing through a record;
(b) extracting the value subfield from the field;
(c) processing multifields;
(d) compactness.

3.19 What are the advantages to using a separate length field for all variable length fields?

3.20 Why does it usually make more sense to use variable length value subfields in a multifield than to have a fixed multifield?

3.21 Compare and contrast the program efficiency of processing a variable multifield using punctuation against other techniques for COBOL or for System/370 (or other) Assembly Language.

3.22 Explain why an absent field is a special case for a multifield.

3.23 Describe another application (besides the invoice or purchase order) for the subrecord principle. Explain your choice for the subrecord.

3.24 A simple record consists of the following fields in this order:
- *ID*—a fixed five-digit field;
- *NAME*—a fixed field, left justified, given name first, twenty-four characters;
- *SEX*—one character;
- *CONT*—a fixed field for up to five four-digit contracts on which he works;
- *CODE*—a one-character employment class code.

   (a) Give one example of this record.

   (b) Give another example, this one showing truncation for *NAME*.

3.25 Use variable length *NAME* fields and convert the examples above, using

   (a) a length subfield,

   (b) $ for a separator.

3.26 Now make *CONT* a multifield for the examples above with

   (a) *CONTN* as a noninclusive number subfield,

   (b) % as subfield separators.

3.27 A savings deposit account transaction record contains these fields:
- *ID*—six-digit account number;
- *DEP*—deposit multifield;
- *WITH*—withdrawal multifield.

   Create a record using % and $ with the following values:

   (a) *ID* = 123456;

   (b) *DEP* = 1.25, 18.75, 25.00;

   (c) *WITH* = 105.80, 77.00, 2.05, 3.75, 8.25, 9.10.

3.28 For Problem 3.27, add *NUM* with two subfields, *DN* and *WN*, and $ to separate subfields in *DEP* and *WITH*.

3.29 For Problem 3.27, use fields *ID, DEPL, DEP, WITHL* and *WITH*, where the second and fourth are *length* multifields starting with *DEPLN*; deposit *length number* and *WITHLN*, *withdrawal length number*; and follow by length subfields.

3.30 Create a master record for Problem 3.27 containing
   (a) *ID*—identification field;
   (b) *BAL*—balance;
   (c) *ACTN*—number of activity subrecords;
   (d) *ACT*—activity subrecord containing *DATE* (six-digit date) and *TRANS* (number of transactions this date);
   (e) *MBAL*—minimum balance this month;
   (f) *CHG*—charge.
   Show a sample record containing three subrecords where *BAL* and *MBAL* are variable length marked with $ and *CHG* is empty.

3.31 Write a BAL program to post the record of Problem 3.27 onto that of Problem 3.30.

3.32 Write a BAL program to post the record of Problem 3.28 onto that of Problem 3.30.

3.33 Write a BAL program to post the record of Probelm 3.29 onto that of Problem 3.30.

3.34 Assume that the subrecord shown in Fig. 3.7.4 is part of an invoice identified by *ID*. Write a BAL program to do two things:
   (a) Find the record for the item in the PARTS file and put it onto it;
   (b) Print a header from data not shown and then one line for each subrecord on a file called OUTPUT. Print one last line for the total.

# CHAPTER 4

# Simple File Use

## 4.1 USE ALTERNATIVES

Assume that we have a file that has been constructed previously. How may we use this file?

- **retrieval**—reference the file to examine information stored there without making any alteration;
- **update**—as individuals in the universe change, alter records to reflect these changes;
- **maintenance**—as individuals enter or leave the universe, create or remove records that represent them.

An application may need one, two, or all three of these uses.

### Retrieval

Figure 4.1.1 illustrates how to use a file for retrieval. On the left is a circle with an X in it, representing the key of a record sought. A string of circles connected by arcs represents the file and any order that may exist in it. The dashed line passing through the file indicates a search. It ends with the sought record. The heavy arrow indicates that the record may be copied.

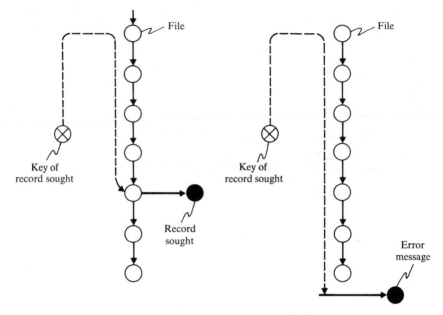

**Figure 4.1.1.** Graph of record retrieval

At the right of the figure, we see an alternative situation where the search of the file is unsuccessful. No record with the desired key is turned up. In this case an error message is produced for the user.

**Update**

Figure 4.1.2 shows update action. The circle with X in it is a transaction record that contains a key and new information about an individual that some record in the file should represent. The dashed line indicates a search of the file for this record. When the record is found, action is taken to transfer (the heavy line) information from the transaction record to the master record so that it is up-to-date.

At the right of Fig. 4.1.2 we see the possibility that there is no correspondent for the transaction record. When the file is searched for the record in the master file represented by the transaction record and none is forthcoming, an error message is produced.

**Birth**

In Fig. 4.1.3 we see a new record represented by a circle with an X in it, which provides information about a new individual who has

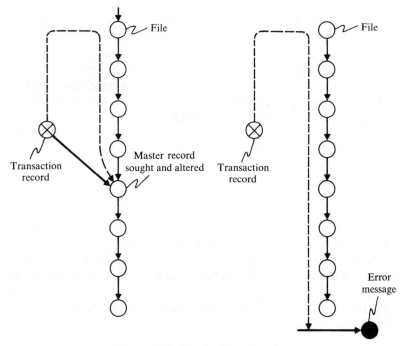

**Figure 4.1.2.** Graph of record update

**Figure 4.1.3.** Graph of record birth

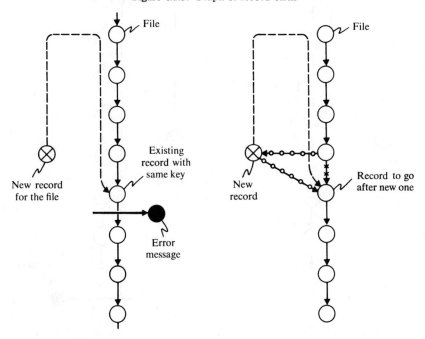

joined our universe. At the left we see a search of the existing file that turns up a record with the same key. Since this is a new individual, there should be no record on file for him, and an error message is created.

At the right of Fig. 4.1.3 a search of the file for the new record reveals that it is absent. The search also determines where in the file the new record should go. Then the new record is entered into the file. The means for doing this graphically is simply by connecting light lines between existing records in the file to show that the record has been incorporated. In actual practice, various techniques can be employed; these will be discussed later.

### Death

In Fig. 4.1.4 we find the key of a record to be deleted represented as a circle with an X in it. The dashed line indicates a file search for a record with that key. When it is found, it is removed from the file. This is shown by the elimination of two light lines and the creation of one new light line.

If the record to be deleted is already absent from the file, the request can be considered fulfilled.

**Figure 4.1.4.** Graph of record death

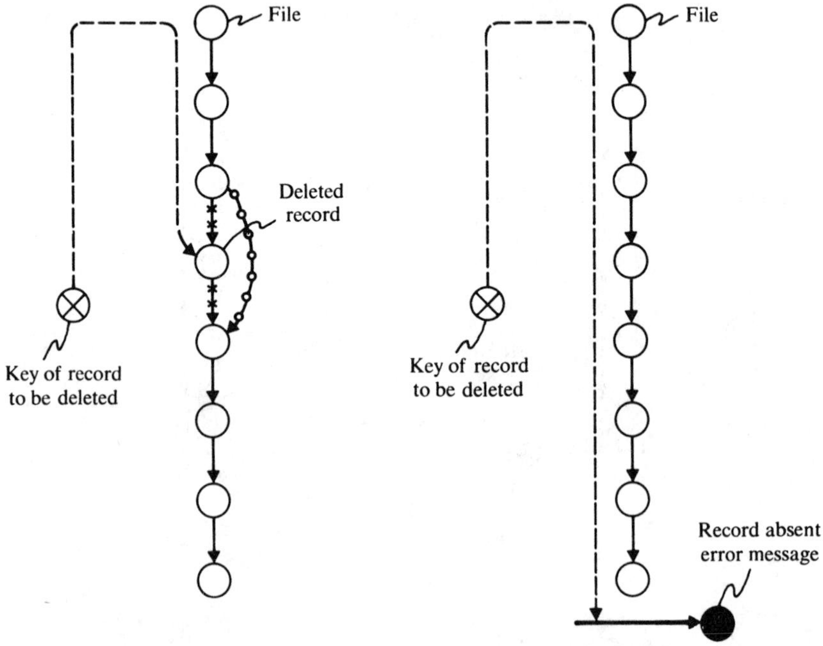

**Statistics**

The file structure that we choose depends on when and how much a file is used in terms of each of the functions described above.

*frequency*     **Frequency** is the number of times a file is used within a given period. We might say that a file is used once a month, twice a week, thrice a day, etc.

*turnover*     Each time the file is used, only a certain percentage of it may be active. The percentage should be known. It is called either **turnover** or **activity ratio**. For instance, for payroll, 95–100% of the file is active. After all, most of our employees do want checks! On the other hand, for a savings account, only 2–3% of our accounts may be active on any one day.

**Combined Outlook**

To choose a file structure we must know the frequency of use and activity ratio with respect to all three functions described above—retrieval, update, and maintenance.

## 4.2 SEQUENTIAL SEARCH

It is apparent from Section 4.1 that retrieval, update, and maintenance all require search. In all cases we search the file for a particular record whose key has been furnished. True, the action is different if the record is present or absent for different needs. Since search is so important, we want to examine different file structures and search methods to see which will produce least search time. At this point we shall contrast only random against ordered file structures. But this contrast shall serve us later in examining other structures.

**The Problem**

We are given a list, **L** and a transaction record, t, which furnishes the key of a record that is sought in **L**. The search is for equal—the key in the transaction record and the record found must be identical. Let us postulate a serial equal search operator *esearch*. Then the action of looking for the key of t in the list, **L**, by examining keys of sequential cells in **L** and putting the **resulting** record in cell R is described symbolically thus:

$$t \ esearch \ \mathbf{L} \rightarrow \mathbf{R} \qquad (4.2.1)$$

Note that *esearch* is a binary operator and t and **L** are, respectively, a prefix and suffix.

If the cells of **L** appear as $L_1$, $L_2$, and so forth, we examine the keys of the records in this order: first $(L_{1K})$, then $(L_{2K})$, etc. Let us now see how the random list and ordered list shape up for the sequential search.

### Random List

For the random list, records are placed into cells sequentially in order of arrival, without any respect to the key of the record. Well, what if the records all arrive at once? Even so, they come in a batch where they appear in the order in which they were created, and we suppose that this too was a random order.

We are interested in the number of looks, $\hat{L}$, needed to find the record we are seeking. Let us now suppose that the record we are seeking is actually present. The number of looks depends on where the record is in the list, and this is a random variable. It is equally likely that the record is in any cell of the list.

It is not important what the number of looks are for any particular list; we really want to know on the average how many looks will be required. Let us define a new variable now, which is called the expected number of looks for finding our record in a random list, assuming it is present, and is symbolized as $\hat{L}_{RP}$, which is given as

$$\hat{L}_{RP} = num \ \mathbf{L}/2 \qquad (4.2.2)$$

Intuitively we can see that it is equally likely that the desired record is at the top or the bottom of the list, and on the average we shall find it in the middle of the list. Therefore the number of looks that we take is the number that will take us halfway into the list, or the average.

Let us call the average number of looks for a random list when the record is absent $\hat{L}_{RA}$. Then we have

$$\hat{L}_{RA} = num \ \mathbf{L} \qquad (4.2.3)$$

That is, if the record is absent, we have to look at all the records in the list before we are sure that it's not one of *them*.

Notice that if we have a loose list, we still have to examine all the cells. Otherwise how would we know that a cell is empty? Thus, (4.2.2) and (4.2.3) hold also for loose lists.

### Ordered List

Suppose now that **L** is ordered and that the record sought is present. The transaction record, t, is chosen at random. Therefore, even though the list is ordered, it is equally likely that the record is in the top or bottom half of the list. On the average, we shall still have to look halfway. Let us use $\hat{L}_{OP}$ to represent the expected number of looks in an ordered list when the record sought is present. Then we have

$$\hat{L}_{OP} = num\ \mathbf{L}/2 \tag{4.2.4}$$

For the same ordered list, what happens when the record sought is absent? In this case, as we look through the ordered list, the key of the records in the cells we are examining provides a clue. As we examine cells, we find increasingly larger keys. When the key in a cell exceeds the key of the record we seek, then we know the record sought is absent. The expected length of such a search depends on the value of the key furnished with t. If this key is low, it should be at the beginning of the list and we shall soon find out that it is absent; if the key is high, the record belongs at the end of the list, and we won't find out until we have passed through most of the list. On the average, we go halfway. Thus, if $\hat{L}_{OA}$ is the expected number of looks in an ordered list when the record is absent, we have

$$\hat{L}_{OA} = num\ \mathbf{L}/2 \tag{4.2.5}$$

### Contrast

It is interesting to note that it makes no difference whether the list is ordered or random if the record for which we search is present in the list—on the average, we go halfway through the list. The difference is evident only when the record we are looking for is absent. Then we must examine the random list exhaustively.

### Find Several Records

Let us now consider the case where we have several records. Let us call this collection of records a transaction file t. Then t contains the keys of a number of records for which we are searching, where the records are found in the list, **L**. We again use the operator *esearch* to designate this search symbolically as

$$\mathbf{t}\ esearch\ \mathbf{L} \tag{4.2.6}$$

*multiple*　　　　One way to approach this task is to take each transaction
*single*　　　　as it arises and look for that key in the list, **L**. Let us call
*searches*　　　the average number of looks required for this method
　　　　　　$L_{tRP}$. For each record the number of looks is given by
(4.2.2). The total number of looks should then be this number multiplied by
the number of records in t. This gives

$$\hat{L}_{tRP} = num \; t * num \; L/2 \qquad (4.2.7)$$

*ordered*　　　We find a definite improvement by ordering both t and L.
*lists*　　　　Let us suppose that both lists are ordered and that the
　　　　　　transaction records in the file **t** are $t_1, t_2$, etc., where the
keys of the records are in increasing order thus:

$$t_1 < t_2 < \cdots < t_n \qquad (4.2.8)$$

We start looking for $t_1$. When it is found, our second search can begin
from there since the key of $t_2$ is known to be higher than $t_1$. A similar statement can be made for $t_3, t_4$, etc. Hence we only need traverse the list **L** once.
We may not even need to go to the very end. If we call $\hat{L}_{tOP}$ the number of
looks required in this case, we have

$$\hat{L}_{tOP} \leq num \; L \qquad (4.2.9)$$

It is important to notice that if some record in t, say $t_j$, contains a key for
which there is no match in **L**, the relation (4.2.9) still holds. After determining
that $t_j$ is absent, we simply continue from $L_m$ to look for $t_{j+1}$, where

$$t_{jK} < (L_{m-1,K}); \qquad t_{jK} > (L_{m,K}) \qquad (4.2.10)$$

## 4.3 THE POSTING PROBLEM

### The Problem

Posting, the largest set of accounting problems handled by
COMPUTERS, is a fairly elementary bookkeeping process. Posting has been
completely automated by the larger commercial enterprises for many years.
Many accounting procedures are actually posting. The four most important
of these are

- payroll;
- accounts receivable;
- accounts payable;
- inventory.

We post to a **master file** (designated here as **m**), which contains all the records about this activity for this company or division, recording the current status of all individuals. There are separate files for each application just as there are separate sets of books for manual bookkeeping. Files are maintained by the COMPUTER; whereas books were maintained by bookkeepers.

Not all the records in a file are active during any posting period, generally. A record is **active** when the individual or item to which it applies has had a change occur in one or more of its attributes. The ratio of the number of active records to the total number of records in the master file is called the **activity ratio**; the elapsed period between each posting is called the **activity period**. The activity ratio and period vary widely from one application to another. For instance, in the banking field, posting is usually done daily with a relatively low activity ratio. For a payroll application, posting may be done weekly, or even monthly, with a very high activity ratio—most employees receive checks.

### The Files

Later in the chapter we examine the organization of files used for posting. Right now we look at the files that are involved and see how records in the file and fields within the records are named.

*old master*    The master file, before it is updated, is the **old master file**, labeled **m**. The $i$th old master record is $m_i$. This record is identified by its key field. We indicate the key field by a subscript K. Thus the key of the $i$th record is $m_{iK}$.

*new master*    The **new master file m'** is the master file after it has been subjected to posting. A record in the file is labeled $m_i'$; its key is $m_{iK}'$.

*transaction*    The **transaction file t** contains one record for each account that has shown activity during this period. The makeup of t—the number of records and their identity—varies from one posting period to the next (different accounts will be active during different posting periods).

*activity*    The **activity file a** is the set of master records that corresponds to records in the transaction file—it contains those master records for which there has been activity during the posting period.

*exception*    Some items may show a kind or amount of activity that is not normal. Management wants to know about this. For instance, an employee may receive a very large paycheck; a bank account may become overdrawn. Management must often individually authorize each activity before it becomes official. The exception file **x** contains records for which activity seems exceptional.

### Operations

Posting consists of a number of operations described as follows:

1. Search—for each transaction record in **t**, we want to find its correspondent in **m**. Section 4.4 is devoted to this operation.
2. Collate—distinguish the active file **a** from the inactive (or passive) file **p**.
3. Update—for each record in **a**, there is a transaction record in **t**. Information from it is transferred by processing some fields of the record in **t** against fields of the record in **a** to form a new record, an updated master in **a'**.
4. Exceptional conditions are noted; when they exist, a new record is created and placed in the exception file, **x**.
5. The updated active records, **a'**, are united with the inactive masters, **p**, to form the new master file, **m'**.

Operations 2 through 5 are discussed in Section 4.5. File maintenance, where the file itself is altered by adding new records or deleting obsolete records, is not a proper aspect of posting. This is discussed in Section 4.6.

### 4.4 INITIAL PROCESSING

#### Single Record

Consider a single record $t$. We indicate searching the master file **m** to find a record in it that has the same key as the transaction record t, using the serial *equal search* operator *esearch* thus:

$$t \; esearch \; \mathbf{m} = a \quad \text{or} \quad \Lambda \tag{4.4.1}$$

The *esearch* operator finds the record a in **m** with the same key as the key of t. That is, $t_K = a_K$. Notice in (4.4.1) that two results are possible: the record a is found or it is absent ($\Lambda$ is the null set symbol showing that a record with the same key as that of t is missing). Then it is possible for the transaction record t to be unmatched! How can this be? Evidently it does not contain a proper key—a key of one of the records in **m**. This might arise from several eventualities.

*errors*     1. t might have been incorrectly copied onto the input medium. This medium is often a punchcard that is key-punched from an original document by an operator who may have made a mistake.
2. The document may have been copied correctly, but it was in error because
    - an external source (such as a supplier for an inventory application) had improperly prepared the document;
    - an internal source (such as an employee for a payroll application) may have improperly prepared the document.
3. The operator did not hit the wrong key but rather mis-read the handwriting of the originator.

The system designer should guard against all these errors. Incorrect transcription can generally be caught by verification as described later.

*check digit*     Origination errors are more difficult to catch. One way to do this is with a check digit. For each record identifier or key a **check digit** is associated by performing some algorithm such as "casting out nines." The check digit and the ID form a **composite identification.**

When this composite identification is submitted, the COMPUTER performs the same algorithm on the ID portion. It should come up with the same check digit. If it does not, one of the digits of the composite identification is incorrect. This kind of error can be caught during the original entry of the transaction, t.

### Procedure

The search procedure to find the corresponding master record depends on the organization of **m**. If **m** is a random file, expect to review half the file if a record is present or the whole file if a record is absent.

Where **m** is an ordered file, the number of looks may be cut in half for an absent item with a simple serial search. It may be reduced more dramatically by a binary search.

### Multirecord File

Let us now examine the case where the transaction file **t** is comprised of a number of transaction records. Now as we pull out the matched master records, they form a separate activity file **a** thus:

$$\textbf{t} \; esearch \; \textbf{m} = \textbf{a} \qquad\qquad (4.4.2)$$

Now the *search equal* operator *esearch* puts into **a** all records in **m** that have the same key as those in **t**. This is shown in Fig. 4.4.1. But not all the records of **t** can be matched. Some of these are rejected. Those rejected transaction records make up a reject file **r**:

$$\mathbf{r} \subset \mathbf{t}; \qquad \mathbf{r} \; esearch \; \mathbf{m} = \Lambda \qquad\qquad (4.4.3)$$

In other words, **r** is the subset of **t** for which there are no records in **m**. For instance, the record with key 7 in Fig. 4.4.1 cannot be matched and is placed in **r**. **a** contains all the matched masters. Of course, 7 is missing.

Finally, there are the master records that are unmatched, the inactive or passive masters **p**. They are defined thus:

$$\mathbf{p} = \mathbf{m} - \mathbf{a}; \qquad \mathbf{m} = \mathbf{a} \; omerge \; \mathbf{p} \qquad\qquad (4.4.4)$$

so that **m** is the ordered union (*omerge*) of **a** and **p**.

### Matching

We could treat the list **t** as though it consisted of single records that we looked up as they arrived. Then **m** could be random or ordered. This procedure would be very inefficient compared to ordering both **t** and **m** because, for each record in **t**, we would have to look, on the average, halfway through **m**. This is independent of where we start in **m**. Hence we would need *num* **t** * *num* **m**/2 looks altogether.

Consider the case where both **t** and **m** are ordered, shown in Fig. 4.4.1. The steps taken in the search procedure are as follows:

1. Get the first (next) record from **t**.
2. Get the first (next) record from **m**.
3. Compare the key of the record t from **t** against that of the record from **m**.
   - If it is smaller, we have gone past the place where the master record should be; **t** is unmatched and it is copied into **r**.
   - If they are equal, we have found an active master. It is placed on **a** and we go back to step 1.
   - If it is greater, this is a passive master p; we have not gone through enough master records. Go back to step 2.

By this procedure, the master list **m** and the transaction list **t** are each examined only once.

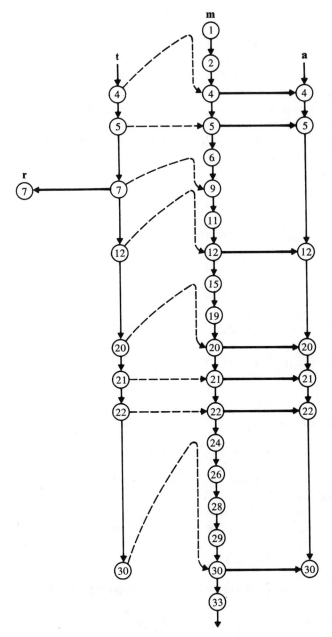

**Figure 4.4.1.** The first action for posting is to create an active master file, **a,** and a reject file, **r.**

### Example

Let us see how the procedure is applied through the example in Fig. 4.4.1. The first transaction record, $t_1$ has a key of 4; the first master record, $m_1$, has a key of 1. Then $m_1$ becomes $p_1$. There is no match. We continue looking at master records until a match takes place. Hereupon, we copy the master record $m_3$ with key 4 to file a to become $a_1$.

After picking up the master record with key 5, we come upon a request for a transaction record $t_3$, with key 7. The master record $m_5$ with key 6 is passed by and the next master record $m_6$ has key 9. Therefore, the transaction record is rejected by copying it onto **r**: $r_1$ is $t_3$; there is no record in **m** or **a** corresponding to $t_3$.

### Transaction Record

The transaction record need contain only fields for those fields in the master that usually change during posting. The only exception is the key—needed for look up—and certain sequence and special fields. The fields are not in the same position as in the master and need not even be the same size. They may be of different size and generally are. The number of active records, *num* **a**—those transaction records for which there are masters—is generally smaller than the transactions and in fact

$$num \ \mathbf{t} = num \ \mathbf{a} + num \ \mathbf{r} \qquad (4.4.5)$$

## 4.5 UPDATING

### Processing

At this point, we have a list of active masters, **a**. Let us suppose that we have a record from this list **a** and the transaction input **t**. We now are going to do some processing so that the record a is updated and is now called a'. This may be noted in operator form thus:

$$\mathbf{t} \ up \ \mathbf{a'} = \mathbf{a'} \qquad (4.5.1)$$

Here *up* is the operation that we perform upon the record a using fields from t; a' is the updated master.

The updating procedure may involve a number of calculations. Just consider a few that might be done in making up a payroll.

- Hours are obtained from t and the rate from a.
- A simple calculation produces the gross pay. This is added to the year-to-date gross in a.
- Each deduction for this period is calculated. Each is subtracted from the period gross pay.
- Each is added to the proper year-to-date deduction.
- We have now found the net pay for the period.
- A year-to-date net pay is also calculated for a'.

**Result**

   The update operation is shown diagrammatically in Fig. 4.5.1. The updated active master, a', is obtained by processing a as defined by the program in the COMPUTER. Besides this, we may have another output file, **n**, referred to as the **normal output file**. It may or may not be present, depending on the application. For instance, for payroll, the normal output is a check and a stub for each individual. For an inventory application, the normal output is a parts order for those items that are getting low. In a banking application, there is no normal output, **n**; we just receive the updated master, **m'**.

In Fig. 4.5.2 we have another output file, **x**, the exception output. It notes records for which exceptional activity has taken place. For instance, a very large salary check, a very low "on-hand" inventory or a very large check being cashed or deposited. This list is printed out for management to examine. It does not affect the process that is going on but does alert management so that normal controls can be instituted if necessary.

**Figure 4.5.1.** The posting operator, $\mu$, processes data from a transaction record, t, onto fields of a matched master record a to create a' and may produce an exception and/or a normal output record, x and/or n.

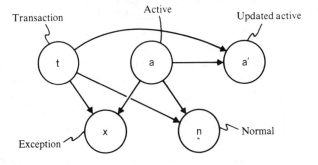

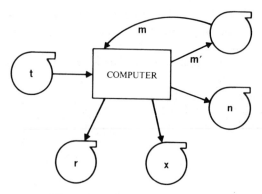

**Figure 4.5.2.** The posting system

### New Master

The new master file, **m′**, consists of the updated active masters, **a′**, and the inactive masters, in order, thus:

$$\mathbf{m}' = \mathbf{a}' \ \textit{omerge} \ \mathbf{p} \tag{4.5.2}$$

Since the old master was in order, we like the new master also to be in order. This leads us to examine the alternatives for creating the new master file **m′**.

With second-generation COMPUTERS, files were kept on magnetic tapes. These do not lend themselves to being updated in place. That is, it is difficult and sometimes impossible to write on a tape from which you have just read. Hence *every* old master record was either altered and copied or simply copied onto another tape. A record for which no activity took place, an inactive master record, p, is copied directly onto the new master file, **m′**, as it is encountered and identified on the old master, **m**. An active record, a, is updated from t **when** a is found on **m** to produce the new master record, a′. Subsequently it is placed on the new master file, **m′**.

The second alternative is to process the records in place. This is possible in third-generation COMPUTERS that use disk packs to store the master file. After an inactive record is read and identified as such, it is left in place undisturbed. As active master records are encountered, they are copied into MEMORY and worked on using a record from t. Upon completion, the new record, a′, is copied into the position that a occupied. After all the records on **m** and t have been examined, the file is said to be **posted**.

Before posting in place begins, it is a good policy to dump the old master, **m**—place it on backup tape in case of difficulty. Several old masters may be kept on hand, as historical masters to reconstruct a file should it get clobbered during posting. When posting is aborted, how could we distinguish updated masters from those that should have been updated?

The third, but less often used, alternative is to create separate ordered files, **a'** and **p**. Then with an additional merge operation the new master file, **m'**, is created.

### Other Files

*reject* The reject file, **r**, which cannot be automatically handled by the system, is printed out to be dealt with manually. The list of rejected transactions is given to a troubleshooter. He uses all methods available to him, including reexamination of source documents, to handle these records. Once the records are correct, the file of corrected rejects, **r'**, can be used in one of several ways:

- It can be run through as though it were a new transaction file for posting.
- It can be entered without sorting by random posting if it is small.
- It can be collated with the transactions **t** for the next period and run at that time.

*others* The normal file, **n**, and the exception file, **x**, are generally produced as printed outputs.

### 4.6 MAINTENANCE

#### Discussion

It is emphasized that posting alters *records in the file but not the file itself*. It is generally necessary, however, to have a method to alter the master file. We distinguish three kinds of file changes:

- Births, **b**, are new records, created and added to the file for new individuals to be added to our universe (books).
- Deaths, **d**, expired records, are deleted.
- Alterations or changes, **c**, are records which alter fields in the master record which are not *normally* altered.

*births* It is easy to see how our files expand. New employees are added to the business, new accounts are buying from us, new merchandise is added to our inventory. Records that carry information about births are different from the transaction records. Really, a birth record is merely a copy of the new master record as it should appear in the master file. Hence, the birth record is generally much larger than the transaction record but the same size as a master record.

*deaths*          Reasons for obsolescence of records are quite similar to those for the creation of new records. But here we must be very careful! For instance, when an employee leaves the concern, his records must be kept on file for some time for tax purposes. We are not so quick to delete this information from the file even though it becomes inactive. An interesting question is what to do when an old employee returns. Do we have the old information on hand and can we reassign him his old employee number?

Again with inventory deletions we may have stock on hand even though it is not part of our regular sales line. Don't we want to sell it, if possible?

A death record, d, need not resemble the master record; it identifies the record to be deleted and provides supplementary information such as expiration data and status.

### Alterations

The novice often misses the distinction between transactions and file alterations. For the payroll example, employee information may change in several fields that are normally unaffected by posting. For instance, his rate of pay, number of deductions, union status, or even her name may change. Making these changes is *not* part of posting, even though it may be done along with normal posting—on the same COMPUTER run.

The alter record c differs from b and d since it must be able to carry information about changes to almost any field of the master record.

### Time of the Change

All three types of maintenance can be made on a single but separate run; they all affect the file structure. It is important to have a different format for each kind of maintenance record to distinguish it from a transaction record and from other maintenance records.

It is possible to intersperse maintenance and transaction records. Then posting and file alteration are done at the same time. Of course, each type of input record must be conspicuously distinguished. For instance, when searching a master file for a birth record, b, we do not expect to find it present there. When it is absent, we do not wish to copy b onto the reject list, r. Instead, we want to enter it onto the new master list, **m'**.

To post and change a file simultaneously is complicated, but as long as the reader distinguishes the two operations, he is now in a position to master the composite activity.

### Summary

- Births are generally the same size as the master records.
- Deaths are often quite small.
- Alter records may come in different sizes and in some cases may be as large as master records.

### PROBLEMS

4.1 For each of the applications below, estimate the activity interval and activity ratio for the function of retrieval, update, and maintainence (six items):
(a) management payroll,
(b) hourly payroll,
(c) accounts receivable,
(d) savings bank,
(e) airlines reservation.

4.2 When a record is sought by key and it is absent from the master file, what action should be taken for the following?
(a) retrieval or reference only,
(b) posting,
(c) birth,
(d) death,
(e) change.

4.3 For batch posting, why should $m$ and $t$ be ordered?

4.4 How does $r$ originate? What are reasons for bad records in $t$? In what way can we protect against them?

4.5 Give an example of a good check digit algorithm. State how it is employed.

4.6 For batch posting:
(a) What is the relation between $m$ and the old $m'$? Why is this so?
(b) Why must $r$ be processed manually?
(c) What is the relation between *num* $t$ and *num* $a$? Why?
(d) What is the relation among *num* $a$, *num* $p$, and *num* $m$, and why?
(e) What relation does $a'$ bear to $a$?

4.7  For the applications of Problem 4.1, state if **x** and **n** are produced and what they are.

4.8  Give a thorough description of how posting might be done for an inventory application to manufactured goods on hand.

4.9  What is file maintenance? Why is it needed? Describe maintenance for a payroll.

4.10  How is posting and maintenance done on the same run? If all the alterations are in the file, **c**, explain how records for birth, death, and change differ. How are **t** and **c** handled for single-run posting and maintenance?

4.11  Apply Problem 4.10 to payroll posting:
(a)  What are the inputs?
(b)  What types of input records are there?
(c)  How do they differ?
(d)  How does **m′** differ from **m**?

4.12  What are the pros and cons of simultaneous posting and maintenance?

4.13  Sometimes the same individual is subject to several changes during one activity. One way to cope with this is to have multiple records in **t** with the *same* key. Examine *all* steps in posting and state what differences in processing are necessary.

# CHAPTER 5

# *Ordered Lists and Sorting*

## 5.1 SEMIDENSE ORDERED LISTS

### Semidensity

A list is called **semidense** when it consists of two parts:

- a dense part, $L_D$, without any holes;
- an empty part, $L_E$, that contains no records at all.

Figure 5.1.1 shows a semidense list in diagrammatic form. The dense part, $L_D$, contains m cells, which are labeled $L_1$ through $L_m$ in the figure. The empty part of the list, $L_E$, contains $n - m$ cells, which are labeled $L_{m+1}$ through $L_n$ in the figure. Each empty cell is marked with @ so that it is clear that the cell is empty when it is encountered.

The dense part of the list is indicated symbolically thus:

$$L_D = \{L_i\}; \qquad (L_{iK}) \neq @ \qquad (i = 1, 2, \ldots, m) \qquad (5.1.1)$$

while the empty part is shown as

$$L_E = \{L_i\}; \qquad (L_{iK}) = @; \qquad (i = m + 1, m + 2, \ldots, n)$$
$$(5.1.2)$$

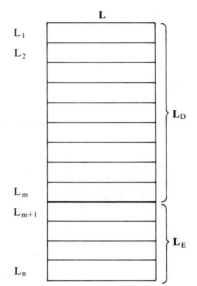

**Figure 5.1.1.** A semidense list

In set theory symbols we have

$$\mathbf{L_D} \cap \mathbf{L_E} = \Lambda; \qquad\qquad \mathbf{L} = \mathbf{L_D} \cup \mathbf{L_E} \qquad (5.1.3)$$

where $\cup$ is the union symbol and $\cap$ is the intersection; that is, $\mathbf{L_D}$ and $\mathbf{L_E}$ are mutually exclusive and together make up $\mathbf{L}$.

It should be noted that *semi* in *semidense* does not mean half. It simply indicates that the list is divided in two parts, which may be unequal, and that these two parts are described with (5.1.1) and (5.1.2). The whole purpose of the semidense list is to provide space for new records so that the file it contains may grow or contract to suit the needs of the user. As this occurs, the boundary between $\mathbf{L_D}$ and $\mathbf{L_E}$ changes, moving down as new records are added or up as old records are deleted.

### Ordered Semidense List

Let us now add the requirement that $\mathbf{L_D}$ is ordered. This may be written symbolically thus:

$$(L_{iK}) < (L_{jK}) \qquad (\text{all } i < j) \qquad (5.1.4)$$

The fact that the list is semidense enables us to use the file sequentially without making special provisions for when holes are encountered. Thus to print out the file it is simply examined from start to finish, where "finish" is determined by the occurrence of the first empty cell (@).

Search of the semidense file can be done sequentially. Then the key in each cell is examined in order of occurrence. Cells are examined as long as the keys so far encountered are smaller than the key sought. For *equal*, the proper record has been found. For *greater*, the desired record is absent since it cannot be in the cells that follow (why?). Provision should be made to stop the search if

- an empty cell is encountered first;
- the list has become dense and we have examined its last cell.

Binary search to be examined later can also be used if the length of the dense portion is known or if we are willing to accept a somewhat less efficient search.

**Appending**

Figure 5.1.2 shows appending to add a record with key frank to an existing list. The first action is to find out where frank belongs. He is greater than ellie but less than gail; hence he should occupy the position occupied by gail. To make this possible, gail and her successors are moved down one cell. This eats up a cell from the empty cells available in $L_E$.

We must be careful about which record we move first; otherwise the action of moving the records may produce an improper copy of the file. It must be remembered that moving destroys the target record but not the one copied. Thus if we move gail to the cell now containing james, he would be destroyed.

**Figure 5.1.2.** A semidense list grows.

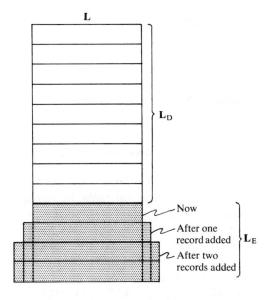

The proper sequence is indicated in Fig. 5.1.2:

- First john is moved to the first available empty cell (1).
- Next james is moved into john's cell (2).
- Finally, gail is moved into james' cell (3).

Now there are two copies of gail; the first copy will be overwritten with frank (4). We now have an ordered file containing one more record than before: the size of the dense portion has increased by one; the size of the empty portion has decreased by one. In other words, we have

$$\text{ellie} < \text{frank} < \text{gail} \qquad (5.1.5)$$

The new sublist is shown in Fig. 5.1.3, which is also used to describe deletion.

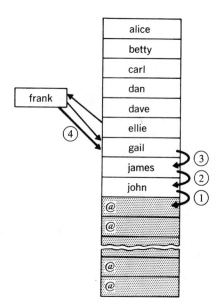

**Figure 5.1.3.** Start to append frank to a semidense list.

**Deletion**

To delete dan from the sublist of Fig. 5.1.4, first find his record. Then, close up ranks by moving everybody backward. Again the sequence of this action is important. First, move dan's successor back to replace dan. After this action we have two copies of dave (1). Next, move dave's successor ellie back to dave's former cell (2), resulting in two copies of ellie. Continue thus until the last record, john, has been moved back into james' position (6). Since we now have two copies of john, the later copy is erased by tagging it with @ (7). This indicates that the cell is reusable but

alice
betty
carl
dan
dave
ellie
frank
gail
james
john
@
@
@
1
2
3
4
5
6
@
7

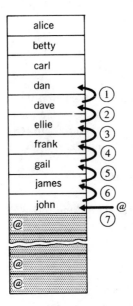

**Figure 5.1.4.** Here frank is added and dan is to be deleted.

john could still be retrieved if we wanted it. Now the new dense portion, $\mathbf{L}'_D$, is smaller by 1 and the new empty portion, $\mathbf{L}'_E$, is greater by 1, thus:

$$num\ \mathbf{L}'_D = num\ \mathbf{L}_D - 1; \qquad num\ \mathbf{L}'_E = num\ \mathbf{L}_E + 1 \qquad (5.1.6)$$

Figure 5.1.5 shows the list after deletion.

**Figure 5.1.5.** Now dan has been deleted.

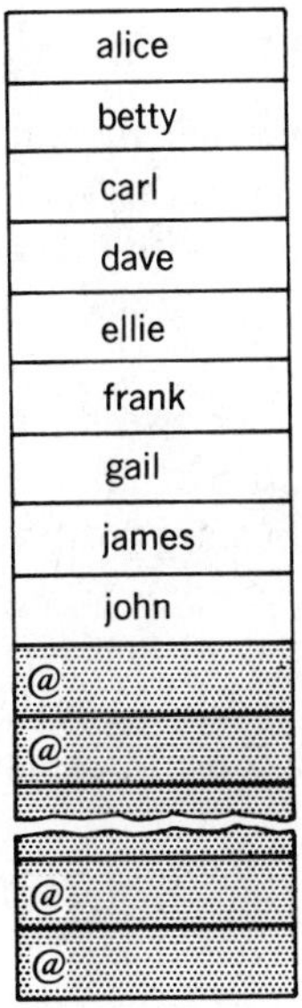

alice
betty
carl
dave
ellie
frank
gail
james
john
@
@
@
@

## 5.2 SORTING IN MEMORY

### Introduction

Sorting creates an ordered list from a disordered file or list. Assume that the records to be sorted are currently in a list called **I**, the input list. What we want to do is come up with a sorted output list **S** where the records are in (ascending) order by key, indicated thus:

$$(S_{iK}) < (S_{jK}) \qquad (\text{all } i < j) \tag{5.2.1}$$

All action on the records found in **I** is done in MEMORY. Whatever method is provided should take into account all possible alternatives for the initial arrangement of records in the list **I**:

- It may be in total disarray.
- Some parts of **I** may be in order.
- **I** may be properly ordered already.
- The list may be ordered but in the wrong direction (in descending order instead of ascending order, for instance).

### Methods Classified

Let us now classify the methods used according to the number and kind of lists provided.

*single list*  It is possible that the sorted list can be created by simply moving about records within **I**. Thus **I** and **S** correspond. We shall still refer to both lists with full knowledge that **S** occupies exactly the same space as **I**.

*two separate lists*  **I** and **S** may be two separate disjoint lists. Records from **I** are moved to **S** so that the resulting list **S** is ordered after the operations are performed.

*auxiliary list*  Some internal sorts resort to constructing an auxiliary list that we shall call **A**. The purpose of **A** is to abstract information from **I**. This information is then used in turn to construct **S** more rapidly.

Often an auxiliary list, **A**, is used in addition to the input **I** and the sorted output **S**.

### Input Restriction

Theoretically all sorting could be done internally provided there is enough MEMORY for the entire list. Practically, this is impossible. Sometimes lists consist of 40 or 50 reels of tape or many disk packs. Current technology does not provide enough internal MEMORY to hold the list in any form.

Suppose that the list is held externally and that this external list is called **L**. Further, all **L** is not simultaneously available to the COMPUTER. Only portions of **L** can be brought in and stored in MEMORY at any given time. We are interested in how a portion of **L** can be put in order. After this has been done, the entire list will be straightened out by successive merging actions that we examine later in the chapter.

Figure 5.2.1 shows the alternatives available for internal sort. Outside the COMPUTER we see the list **L**. It is brought into the area **I** (1). Sorting being done in this area is shown diagrammatically (2). If records are moved from **I** to **S** to provide the rearranging, this is also shown (3). The final alternative is to create an auxiliary list (4) that guides records from **I** to **S**.

We now introduce conceptually two of the sorts that will be discussed in more detail in Section 5.3.

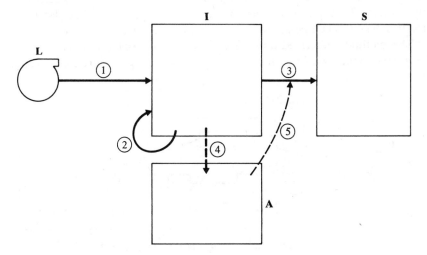

**Figure 5.2.1.** Alternatives for the internal sort

### Insertion Sort

For this sort, **I** is kept as a semidense list. As a record is brought from **L**, its place is found in **I** and the dense portion of the list reshuffled to accommodate the new record. This action is continued until **I** is full. At that point, the records in **I** are ordered, so it may now be referred to as **S**.

### Selection

For selection, load **I** serially with records from **L** until **I** is full. Next examine records in **I**, looking for the smallest one, which is then placed at the beginning of **S**. Then, return to **I** and find the next smallest record, placing it in the next position in **S**. At the end of this activity the records from **I** are now all in **S** but in the proper order.

### Replacement

Some internal sorting methods lend themselves to replacement, which is illustrated in Fig. 5.2.2. For instance, selection by replacement is done as explained above and shown in the figure. But as a record is moved from **I** to **S**, a new one is brought in from **L**. Thus the first record is selected from **I** and copied into **S** (1), and a new one is brought into the spot in **I** occupied by the selected record (2). A similar action occurs for the next record, which is selected from **I**, copied into **S** (3), and then replaced by a record from **L**. Action continues thus until **S** is full. We shall see later that for this design **S** can be approximately twice as large as **I**.

The methods described above serve only to convey sort concepts. The user would employ a sort package that might use a much more sophisticated technique.

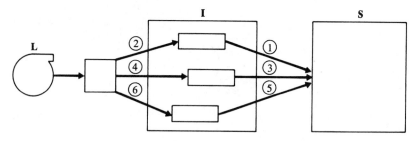

**Figure 5.2.2.** Replacement sorting

## 5.3 SORT DETAILS

### Insertion

For sort by insertion, records are coming into the COM-PUTER from the external list, **L**, and they are presented to the program in arrival order, whatever that might be. The output for the sort is the semidense ordered list, S. The size of the list is predetermined in the program or pre-initialized by the user, as described in Section 5.5. When the action starts, the semidense list is empty—it contains no records—there is no dense part.

Sorting begins as the first record from L is placed at the first position in S. The second record is then examined and is placed either after the first record, if it has a higher key, or the first record is moved down so that the second record may precede it.

Sorting proceeds as though we were appending a new record into a semi-dense list, for that is truly what we are doing. As each new record is encountered, its place is found in $S_D$. The records from this spot on are moved forward to make a cell available for the new record while maintaining order in the dense portion.

We continue to bring in new records until S is full. For fixed-size records, S will contain a predictable number of records. For variable length records, we could only tell that S is full when the next record from L will not fit in the space now available in $S_E$.

During the total sort, when S becomes full, it is written onto the output medium and S becomes available again to produce a new input stream.

### Selection

For sorting by selection, the input area I is filled with records. The least of these will be selected for output, then the least of the remaining records, and so forth.

The selection sort is illustrated in Fig. 5.3.1 through 5.3.3. In these figures, three registers hold particular quantities, designated as follows:

IKR contains the key of the smallest record encountered so far.

IPR contains a pointer to the call address for the record for which the key is now in IKR.

**Figure 5.3.1.** Sort by selection after excusing the first three records the first time.

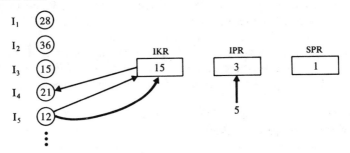

SPR, the SORT POINTER REGISTER, points to the place where the output record will go in **S**.

***initialize***    To begin the search, the first record in **I** is examined and its key extracted and placed in IKR. The location of this record or its ordinal number is entered into IPR:

$$(I_{iK}) \longrightarrow \text{IKR}; \qquad 1 \longrightarrow \text{IPR} \tag{5.3.1}$$

Search continues by looking at the next record to determine if it has a key smaller than that held in IKR. If so, that key is placed into IKR and the position of the record into IPR. The records in **I** are examined exhaustively thus. At the point where we are examining the jth record $I_j$, the action performed is indicated with

$$(I_{jK}) < \text{IKR}; \qquad (I_{jK}) \longrightarrow \text{IKR}; \qquad j \longrightarrow \text{IPR} \tag{5.3.2}$$

If the key in the jth cell is larger than the key in IKR, do nothing.

The search is over when all the records in **I** have been examined. At this point we have

$$(\text{IKR}) \leq (I_{jK}) \qquad (\text{all } j) \tag{5.3.3}$$

That is, the key in IKR is less than or equal to the key of any record in **I**.

***example***    In Fig. 5.3.1 we see the contents of IKR and IPR after we have examined the first three records of **I**:

- IKR equals 15, the key of $I_3$.
- IPR contains 3 since that is the ordinal position of $I_3$.
- SPR contains 1 since the first record has not been placed into **S** yet.

Note that, in an actual sort, IPR will probably contain the *absolute location* of $I_3$.

The key of the record at $I_4$ is 21. Since 21 is larger than 15, the three registers are not affected. Upon reaching $I_5$ we note that its key is 12, which is less than 15. Consequently, its key is placed into IKR and its position, 5, is entered into IPR, as shown by the heavy arcs in the figure.

***transfer***    Upon completion of the search the IKR contains the key of the least record and the IPR contains its location, as demonstrated in Fig. 5.3.2. We are now ready to transfer the record from **I** to **S**:

- IPR indicates the source of the record, $I_{10}$.
- SPR indicates the destination, $S_2$.

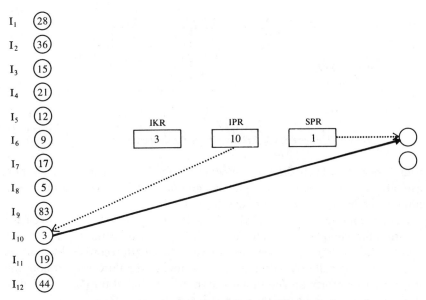

**Figure 5.3.2.** Sort by selection after choosing the first output record.

• The solid line in the figure shows the transfer of the record to the empty cell.

This is indicated symbolically as

$$((\text{IPR})) \longrightarrow (\text{SPR}) \tag{5.3.4}$$

where IPR is a register, (IPR) is its contents, a pointer, and ((IPR)) is the record pointed to.

*update*    If we were to perform another search of **I** as described above, we would come up with identically the same results. We must eliminate from consideration the record that we have just copied from **I**—it still exists there too. There is no need to destroy the entire record in **I**. We write over the key some indicator that will make, on comparison, all other other keys seem larger.

I use the symbol $\infty$ to indicate a key larger than any other key. This key is written into the record we have just copied and is pointed to by IPR, as indicated thus:

$$\infty \longrightarrow (\text{IPR}) \tag{5.3.5}$$

In Fig. 5.3.3 we see what happens after the record at $I_{10}$ with key 3 was chosen for output. First, $\infty$ is written at $I_{10}$ to replace the key originally

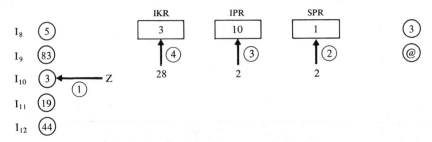

**Figure 5.3.3.** Entering Z for the chosen record

found there (1). Then the SPR is updated from 1 to 2 to point to the next available space in **S** (2). Finally, the IPR and IKR are initialized to the beginning of the list **I**, where search may begin (3, 4).

Action continues thus until the input list **I** is exhausted. For fixed-size records, the completion of the sort is determined by counting the records copied. This technique can be used for variable length records if the number of records originally placed in **I** has been noted. The third way to note this is simply to perform an additional search, which should turn up a key of ∞ in the IKR, which indicates the completion of the sort, thus:

$$(\text{IKR}) = \infty: \quad stop \tag{5.3.6}$$

### Selection and Exchange

Selection and exchange is quite similar to simple selection, except that it results in rearrangement of the original input list, **I**. This offers several advantages:

- No extra space is required for **S**.
- Instead of keeping a copied record and altering its key, the record is placed elsewhere in the list where it will no longer be considered.
- Instead of examining the whole list for each search, only the unordered part is examined.

During the sorting procedure, the original input list, **I**, is considered to consist of two parts:

- $I_S$ is the sorted part at the top.
- $I_U$ is the unsorted part at the bottom.

This is much like the semidense list; we might call this the semiordered list. As sorting proceeds, the ordered portion grows in length and the unordered portion decreases until finally all the records are in the ordered portion.

***initialization*** Keep the same register designations. The first search proceeds here as described for the simple selection sort. Upon completion, the IKR, IPR, and SPR contain, respectively, the key and position of the least record and a pointer to the top of the *input list*, as shown in Fig. 5.3.4.

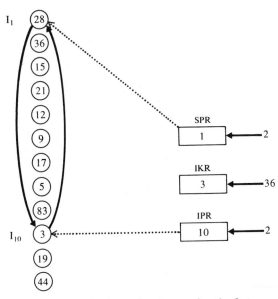

**Figure 5.3.4.** Selection and exchange after the first pass

Now, instead of moving the selected record, it is exchanged with the record at the top of the list. In the example shown in Fig. 5.3.4, the record with the least key of 3 is found at $I_{10}$, pointed to by IPR. An exchange takes place between $I_1$ and $I_{10}$ so that the least record is put at the top and $I_{10}$ contains the record with key 28. The SPR and the IPR contain the address of the records that are to be interchanged.

Note carefully that the records are to be *interchanged*, transferred without interfering with each other (1). An intermediate location (say TR) may be used since copying $I_1$ to $I_{10}$ or vice versa will clobber the other record. Then we have

$$((\text{IPR})) \longrightarrow (\text{SPR}); \quad ((\text{SPR})) \longrightarrow (\text{IPR}) \qquad (5.3.7)$$

***reset*** Now that the records have been interchanged, $I_U$ consists of only eleven records. Reinitialization consists of resetting both SPR (2) and IPR (3) to the top of the unordered list, $I_U$, and moving the key of the first record of $I_U$ into IKR (4), as shown in Fig. 5.3.4. This is shown as

$$(\text{SPR}) + 1 \longrightarrow \text{SPR, IPR}; \qquad ((\text{IPR}))_{\text{K}} \longrightarrow \text{IKR} \qquad (5.3.8)$$

Comparisons for the ten remaining records now take place. The result of this next set of comparisons is shown in Fig. 5.3.5. The third set of comparisons is illustrated in Fig. 5.3.6.

**Figure 5.3.5.** Selection and exchange after the second pass

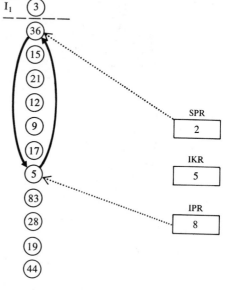

**Figure 5.3.6.** Selection and exchange after the third pass

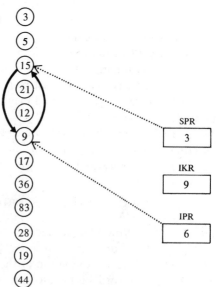

***termination*** Action continues thus until all records are in their proper place, determined either by counting the number of times the list is examined or by noting the position of the SPR as a new comparison is started.

### Selection with Replacement

Many sorts can benefit by using replacement. Selection is not considerably improved by replacement but it is easy to explain the concept using selection as an example.

For replacement, three lists are defined:

- **L** is the original list as it exists on the auxiliary medium.
- **I** is the unordered list in MEMORY that is to be sorted.
- **S** is the sorted output list.

For replacement, as records are copied from **I** into **S**, the cell previously occupied by the copied record is used to hold a new incoming record from **L**.

Replacement is most beneficial when applied to sorting techniques that use auxiliary lists, such as the quadratic sort or the tournament sort, described elsewhere.† For replacement with selection, the selected record is replaced by an incoming record and selection proceeds as before.

When selection and replacement are employed from a given size of input list **I**, we can develop a larger output list. Some of the new records that replace those put in **S** will eventually be sorted into this output list **S**. Others, smaller than the last record put in **S**, will not be used and we eventually run out of records for **S**. It has been shown mathematically that the output list will be approximately twice as large as the input list.

***example*** Let us assume that the output list **S** has already received i records and that the last record has come from $I_j$. Replacement selection requires that this cell be filled with the next available record from outside (**L**). If **I** contained n records initially, then the new record being brought in will bear the subscript n + 1, so that we have

$$L_{n+1} \rightarrow I_j \tag{5.3.9}$$

***exception*** There is one exception to the selection procedure that is now employed. First, it is clear that a record in **I** need no longer be excluded from consideration by inserting $\infty$ in the key position of its cell in **I** after it has been selected for **S** since such a record has now been replaced by a new incoming record. Hence the entire list **I** will be examined.

† Ivan Flores, *Computer Sorting*, Prentice-Hall, Englewood Cliffs, N.J., 1969.

We do not simply want to find the *least* record currently in **I**, however. We have a new qualification for output list eligibility. If we chose the least record, it might be smaller than the last record that we put in **S**. Therefore our search of **I** is for *the least record that is greater than the last record placed in S*.

Again, we initialize our search as for simple selection:

$$(I_{1K}) \rightarrow \text{IKR}; \quad 1 \rightarrow \text{IPR}; \quad j = 2 \tag{5.3.10}$$

Now when we look at the jth record in **I**, $I_j$, we make sure that two conditions hold before we update the registers concerned:

- Its key should be less than the one found in IKR.
- Its key should be greater than the key of the last record placed in output at $S_i$:

$$(I_{jK}) < (\text{IKR})$$

and

$$(I_{jK}) > (S_{iK}): (I_{jK}) \rightarrow \text{IKR}; \quad j \rightarrow \text{IPR}; \quad j + 1 \rightarrow j \tag{5.3.11}$$

If this isn't so, we simply go on to the next record in **I**:

$$(I_{jK}) > \text{IPR} \quad or \quad (I_{jK}) < (S_{iK}): j + 1 \rightarrow j \tag{5.3.12}$$

***example***      In Fig. 5.3.7 we see an example. Previously, $I_6$ contained a record with key 9. This record became $S_3$, the third output record. $I_6$ now contains a record with key 7, the replacement record. Let us now examine records of **I**. When we come to $I_4$, it has a key of 21, which is greater than the key of 15 contained in IKR; we ignore it since we have encountered a smaller record. At $I_5$ we find a key of 12, which is smaller than 15 now in the IKR. Since it also is greater than 9 in $S_i$, we enter the new information into IKR and IPR.

**Figure 5.3.7.** Selection with replacement where a small key record replaced a larger output record.

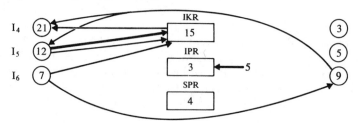

Now we encounter the key of 7 in $I_6$. It is certainly less than the key of 12 now in the IKR, but it is also less than the key of 9 for the record last selected, so it is ineligible for the list S. That is, records for S are in ascending order and though 7 at $I_6$ is a correct number, it doesn't belong after 9.

Note that if $I_6$ contained a record with key 11, it would be eligible for output and eventually would be found in the output list S.

**termination**    Sorting and replacement continue as long as records are produced for the output list S. When all the replacement records found in **I** are *less* than the last reocrd in S, we have no more items to add to that list and it is therefore complete:

$$(I_{jK}) > (S_{mK}) \quad \text{(all j:} \quad stop) \tag{5.3.13}$$

**I** contains a whole new batch of records, however, all of which are less than the last one in S. Let us get rid of (output) the current list now in S and start anew. Now find the lowest record in the new batch and start a new output list.

Eventually, when a record is placed on the output list and we go to **L** for a new record, we find that there are none. This requires that the empty cell in **I** be so tagged. Sorting of the records in **I** will continue *without replacement* until **I** too is exhausted. Then all the required sorted lists have been produced.

## 5.4 MERGING

### Need

The internal sort creates a list that contains several ordered subfiles. We want a single ordered file. The MEMORY available in the COMPUTER limits the size of the ordered file that can be produced by an internal sort. We accept this restriction and produce files containing several ordered subfiles. Now the problem is to combine two or more ordered subfiles to make a single ordered subfile. This action is called **merging**. If we repeatedly apply merging to files of ordered subfiles obtained through internal sorting, we should eventually obtain a single file that is ordered.

In Fig. 5.4.1 we begin with two completely ordered files **a** and **b** from which we build a single ordered file **c**. In the figure and the descriptions that follow, each record is treated as though it were only a key. Actually the entire record is moved.

After presenting the merging of two fully ordered files, we investigate general merging of two files, each of which consists of several ordered subfiles. Later, sorting a completely disordered file is examined.

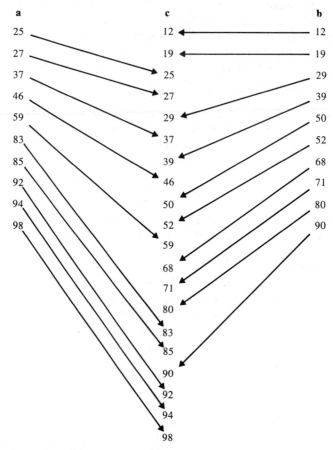

**Figure 5.4.1.** Merging ordered files **a** and **b** to form **c**

### Merging Two Fully Ordered Files

To merge two ordered files, **a** and **b**, into a single ordered file, **c**, we could begin at the top record (smallest key) or bottom record (greatest key) of both files. Let us choose the first alternative. What rules apply to the records?

*first*     In Fig. 5.4.2 we look at the order relation as it applies to the two top records. The arrow from $b_1$ to $a_1$ shows that $b_1$ (12) is smaller than $a_1$ (25); $b_1$ is therefore placed at the top of our new file **c** to become $c_1$. The heavy line indicates the movement of the record $b_1$ from **b** to become $c_1$ of **c**.

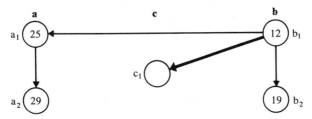

**Figure 5.4.2.** Choosing the first record for **c**

*general* Suppose we are examining the ith record, $a_i$, of the file **a** and the jth record, $b_j$, of **b**. The arc in Fig. 5.4.3 shows the order relation of $a_i$ and $b_j$ to each other and also with respect to the most recent addition to the merged list, $c_{i+j-2}$. In this well-behaved situation, both records are larger than the last in the merged file. (Why?) Therefore, we choose the smaller of $a_i$ and $b_j$ and make it the next addition, $c_{i+j-1}$, to **c**.

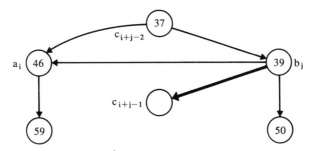

**Figure 5.4.3.** Choosing a general record for **c**

*one file* Sooner or later one of the files is used up. In Fig. 5.4.4 we
*exhausted* have placed the bottom record from **b**, $b_m$, on the merged file. There are no more items left on **b**; therefore, all the remaining records on a are placed in the same order on **c**. This is called **roll-out**. They all have keys greater than that of $b_m$. (Why?)

### Files of Ordered Subfiles

Two files of ordered subfiles **a** and **b** are merged to **c** in Fig. 5.4.5. A record in each file is seen to have two arcs emanating from it and none entering it. Each is the top record ($a_5$ and $b_5$) of the next subfile of **a** and **b**, respectively. These records, $a_5$ and $b_5$, are less than the records above *and* below them (which belong to different subfiles). These are the step-down records: a **step-down** distinguishes ordered subfiles within the file; it is noted when the key of a record is *smaller* than that of the preceding record.

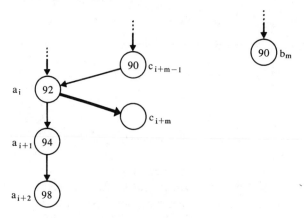

**Figure 5.4.4.** Rollout

**Figure 5.4.5.** Merged with files of ordered subfiles

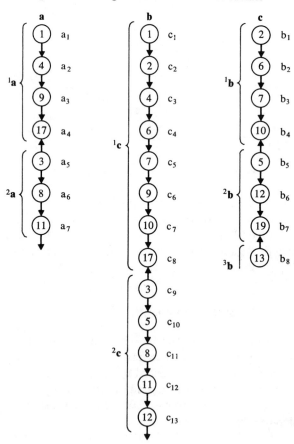

In merging of two files of subfiles, four cases arise, discussed below in terms of the number of step-downs that have occurred in creating **a**.

***no***
***step-downs***

Figure 5.4.6 illustrates a pair of subfiles where no step-down occurred. Both new records, $a_i$ (17) and $b_j$ (10), are greater than the most recent addition to the merged file, $c_{i+j-2}$. We demonstrate with $a_4$, $b_4$, and $c_6$ of Fig. 5.4.5. In general the lesser of $a_i$ and $b_j$ is entered as the next record on the merged file as $c_{i+j-1}$. (Why?) In the illustration, $b_4$ (10) is the smaller and becomes $c_7$.

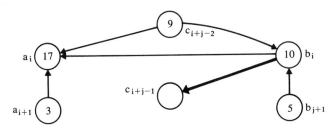

**Figure 5.4.6.** Choosing a general record for **c**

***single***
***step-down***

In Fig. 5.4.7 a single step-down has just occurred in $b_j$. We demonstrate with $a_4$, $b_5$, and $c_7$ of Fig. 5.4.5. Hence the current record $b_5$ (5) is *less* than the last, $b_4$ (10); $b_5$ is hence less than the last **c** record, $c_7$. Further, $a_4$ (17) is larger than the record $c_7$. A step-down in **b**, however, inhibits the use of **b** records until **a**

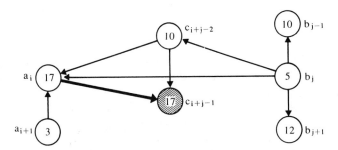

**Figure 5.4.7.** Single stepdown

also has a step-down. Therefore, $a_4$ is chosen for **c** as long as $a_4$ is larger than $c_7$. We now empty the ordered subfile of the remaining records onto the output subfile. The top of the next ordered subfile of **a** is distinguished by a step-down record.

If the single step-down occurred on **a**, we would take further records for **c** from **b** until a step-down occurs in **b**.

***double***
***step-down***

Figure 5.4.8 illustrates a double step-down. We have now merged two subfiles, one each from **a** and **b**, and produced a single ordered subfile on **c**. The last record on this merged subfile (17) is necessarily larger than either new record, $a_i$ or $b_j$, in **a** (3) and **b** (5). A double step-down has occurred. For the next merged subfile we choose the smaller of the two records presented (3). The status now reverts to *no step-downs*.

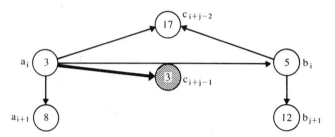

**Figure 5.4.8.** Double stepdown

***terminal***
***rollout***

As demonstrated in Fig. 5.4.4, after the *last* record in **b** is placed in the merged file, *all* records from **a** go to **c**—**rollout**. This continues regardless of whether step-downs occur on **a** (because there may be more than one ordered subfile remaining on **a**). Although rollout of **a** has been demonstrated, it is clear that some merges will call for a rollout of **b** instead.

### Multiway Merge

We have seen how two lists containing ordered sublists can be merged using a two-way merge. If we make the program more complicated, we can handle several lists of ordered sublists. The number of lists provided as input determines the number of **ways** provided in the merge.

An example of the three-way merge is shown in Fig. 5.4.9. **a**, **b**, and **c** are files that contain ordered subfiles. These will be merged to form a single file containing ordered subfiles. That file is called **d** in the figure.

***procedure***

The procedure employed to select the next record for the output file is exemplified in Fig. 5.4.10 for the three-way merge. The principle here can be extended further for the four-way, five-way, etc., merge. The input files to be used, **a**, **b**, and **c**, are the same as those presented for Fig. 5.4.9. The input records are subscripted according to their

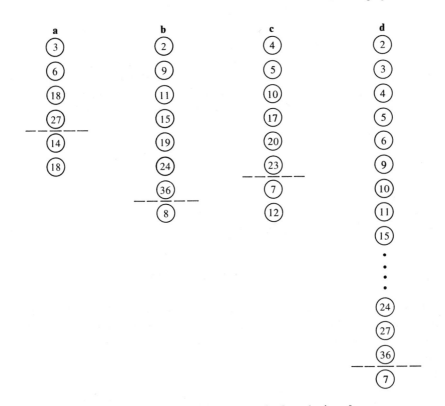

Figure 5.4.9. Three-way merge of **a**, **b**, and **c** into **d**

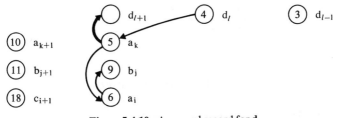

Figure 5.4.10. A general record for **d**

ordinal occurrence in the file to which they belong. We are now examining $a_i$, $b_j$, and $c_k$. The most recent output record is labeled $d_\ell$. The next item will be $d_{\ell+1}$.

Figure 5.4.10 shows how we examine items in the three files that have been laid down so that they appear horizontally instead of vertically. The top records in the files are compared to find the smallest. In the figure, $c_k$ contains the smallest and will be placed at $d_{\ell+1}$. This is possible since no step-down has occurred and all records are larger than the last output record.

***step-down***    A step-down occurs on a file when we encounter a smaller record, indicating that we are entering a new sublist. A single step-down has occurred on **c** in Fig. 5.4.11. The record at $c_k$ has a key of 7, less than that at $d_\ell$, which has a key of 23. For a single step-down, hereafter we shall effectively be doing a two-way merge, considering records only from **a** and **b**. In the figure, $b_j$ with key 24 is smaller than $a_i$ with key 27 and hence is selected for output.

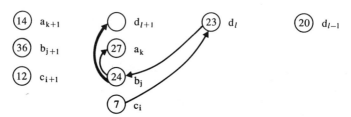

**Figure 5.4.11.** Single stepdown

In Fig. 5.4.12 we see the case of a double step-down. A step-down has occurred on both **a** and **c**. In this case a copy action is performed from **b**. This is true since the records in the other file are smaller than the last record (36) for output.

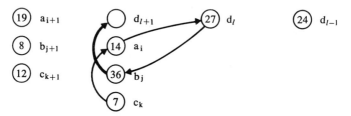

**Figure 5.4.12.** Double stepdown

When a triple step-down occurs for the three-way merge, we are restored to the *no step-down* condition. Merging continues on a three-way basis until the first step-down occurs, and so forth.

***rollout***    When a file expires, an end-of-file designator is encountered so indicating. Action continues using the remaining files. The power of the merge decreases accordingly. For the three-way merge, two-way merge begins the action called **rollout**. When only one file is left, the remaining records on it are copied for output. Its completion signals the end of the merge operation.

## 5.5 THE SORT UTILITY

### What Is It?

The sort utility is a rather large program. Most manufacturers have at least one sort utility that they supply free to the user. They may have other and better sort utilities that they provide to the user on a fee basis, rental or purchase. Competitive software houses also lease or sell sort utilities that are tailored to particular COMPUTER types.

The sort utility provides an ordered list of records from a disordered list supplied by the user. It should do this with efficiency and reliability and ensure the integrity of the user's data.

Since the utility contains both a sort and a merge phase, some utilities enable the user to use either one independently. The extensive sort utility allows the user to name compound sort keys, so that the sorting is done on mixed key fields. The utility allows various combinations of DEVICES for work space to be used during the merge operation. It may further provide a choice of different kinds of merges to optimize that activity.

*phases*　　　　The activity of the utility can be divided into four phases, as follows:

1. The control and generate phase determines the user's needs and tailors code to meet these needs.
2. The internal sort creates lists of ordered subfiles and distributes them to work DEVICES.
3. The merge provides several different steps during which merging is done.
4. The final merge places a single list of data onto the medium prescribed by the user and in the proper order.

### Control and Generate Phase

The first phase of activity for the sort utility is to obtain instructions from the user about the data and what to do with it. Then the availability of the DEVICES provided is verified and work areas are set up on each medium. Now the code that is to perform the internal sort is generated and tables are set up to locate the fields that make up the key. Finally, the proper merge technique is selected and a distribution procedure is determined so that after the sort phase is complete, merge can begin properly.

Control statements in the job stream convey the user's specification of the data and the techniques desired. The job stream contains the **job control language** statements (**JCL**), which describe

- the overall job;
- each step that comprises the job;
- every file to be used on each step;
- control statements for sorts and other utilities when employed.

To see the kind of information required as control statements, let us take as an example the sort utility provided free by IBM for their OS System. Incidentally, the control statements used there are almost identical to those used by their program product, which is supplied at a fee to the user.

Let us consider an example. The first statement supplied is.

SORT FIELDS = (10, 5, CH, A, 20, 3, PD, D), SIZE = E10000    (5.5.1)

The key word **SORT FIELDS** indicates that we are describing the key fields to be used during the sort. Within parentheses we find eight specifications. The first two numbers indicate the number of the byte at which the key begins and the key length (i.e., the first key begins at byte 10 and is five bytes long). The next parameter indicates the kind of data that appears in this field (i.e., CH indicates that character data is found in the first key field). The fourth parameter indicates the order with which the file is to be sorted with respect to this key (i.e., A indicates ascending order for the output file).

The next quadruple describes the next key segment in the same fashion (i.e., the next key segment begins at byte 20, is three bytes long, consists of packed data, and orders records in descending order).

The last parameter in the statement indicates the size of the file by giving the number of records. If the size is an estimate, the number is prefixed by E.

The next statement gives further information about the record:

RECORD TYPE = F, LENGTH = 80                        (5.5.2)

The record type may be fixed (F) or variable (V). LENGTH is given in bytes and indicates the actual record for fixed length records or the maximum record size for variable length records.

One further specification that we allow to the user is one or more hooks by which a user-written routine can gain control during a sort. A **hook** is an optional exit from the package to a user-supplied routine. By studying the sort manual, the user can determine when given activities have been completed. He may request control at specified times to do such things as take an accurate count of records and determine when invalid records have been noted. A routine that he has written must be available at sort time in a library or in the input stream. He indicates his request thus:

$$\text{MODS E35} = \text{(CALC, 800, USERLIB)} \qquad (5.5.3)$$

Here E35 is a specified point in the sort activity. CALC is the name of his exit routine, 800 is its length, and it may be located by the sort utility in a library called USERLIB.

### Sort and Distribute

The sort utility often provides a particular sorting technique that is used regardless of the characteristics of the list furnished. In some cases, the utility will optimize the sort method according to the position and length of the key and its complexity and taking other file characteristics into account. Generally, as shown in Fig. 5.5.1, the input list comes from a single medium, however, it is possible to specify several different files for input. In this case, the files are entered in the order mentioned in the JCL. It is the JCL file definition statements that declare which DEVICES and media contain the input files. For OS this is the //DD or data definition statement. Similarly, the work files are defined in the JCL.

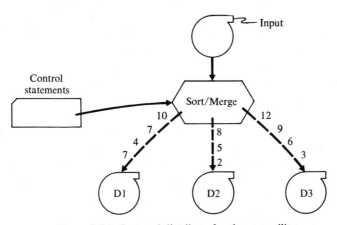

**Figure 5.5.1.** Sort and distribute for the sort utility

The sort phase creates ordered subfiles and puts them on work files where they are further acted upon by the merge. Obviously if all the ordered subfiles are put into the same list, it would be very difficult to merge them.

The sort utility distributes ordered subfiles to different lists. The sequence in which this is done depends on the kind of merge and its complexity. For the balanced multiway merge, which will be described shortly, files are distributed equally to each DEVICE. Thus in Fig. 5.5.1 as each subfile is produced, it is sent to a different DEVICE. When each DEVICE has been furnished exactly one subfile, we start with the first DEVICE again and then distribute the next subfile to the next DEVICE. We may call this **round robin distribution.**

In the figure the first subfile goes to DEVICE 1 (D1); the next, numbered 2, to DEVICE 2 (D2); the third, to DEVICE 3 (D3). Since there are only three DEVICES, the fourth subfile goes to DEVICE 1; the fifth, to DEVICE 2; and so forth.

### Balanced Merge

A balanced merge is so-called because an equal number of input and output DEVICES are provided for the merge. The **order of the merge** is defined as the number of input or output DEVICES available. A three-way merge is illustrated in Fig. 5.5.2.

To begin the merge, the OUTPUT DEVICES for the sort and distribute phase become the INPUT DEVICES for the merge (i.e., D1, D2, and D3 in the figure). The same number of output DEVICES are required (D4, D5, and D6 in the figure). The merge reads records from each input DEVICE and combines the subfiles presented into a single subfile. Thus subfile 1 from DEVICE 1, subfile 2 from DEVICE 2, and subfile 3 from DEVICE 3 are combined into a subfile that we shall call 123, which is placed on DEVICE 4 shown in the figure. New subfiles 4, 5, and 6 are then worked on. If we place these on the same DEVICE, further merges would be difficult if not impossible. Hence each merge requires a distribution activity to place merged subfiles on alternate DEVICES according to some discipline. In the figure, subfile 456 (fabricated from subfiles 4, 5, and 6) is placed on DEVICE 5. Similarly, 789 (made from input subfiles 7, 8, and 9) is placed on DEVICE 6. Subfile 101112 will go to DEVICE 4 as we start the round robin again.

As we complete this first merge, we find that each list on the output DEVICES contains ordered subfiles. There are fewer subfiles and each subfile is longer. Generally the factor involved is the order of the merge. Thus for the three-way merge illustrated, each subfile is approximately three times as long as the ones provided at the input of the merge; whereas there are approximately one-third as many subfiles in each list.

On completion of the first merge, the output DEVICES contain lists of subfiles that have duplicates of the records found in the lists supplied as input. The input lists are no longer needed. Hence the media on the original input DEVICES can be reused to receive subfiles produced by the next or second merge. The first action of the second merge is to rename INPUT DEVICES as OUTPUT DEVICES, and the OUTPUT DEVICES are then named as the new INPUT DEVICES. According to the media used, reinitialization may be required—that is, for tapes, for instance, the tape drives may be rewound.

**Figure 5.5.2.** The balanced merge

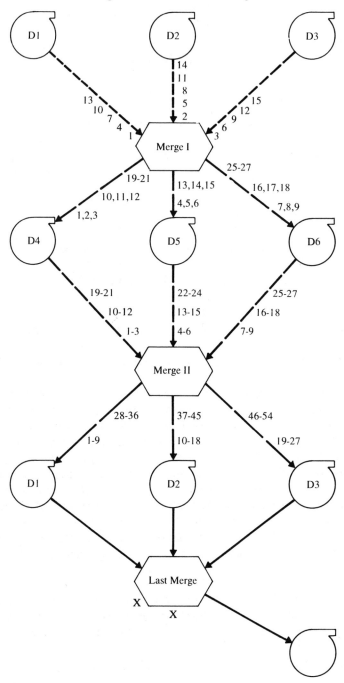

Figure 5.5.2 also shows the action required during the second merge of the three-way balanced merge described earlier. The merge begins by taking the subfiles 123, 456, and 789 into a new subfile named 1–9, which is distributed to D1. The next merged subfile, 10–18, ges to D2, and the third one, 19–27, goes to D3. When the fourth one, 28–36, is created, it is placed on D1 as should be apparent.

The finish of the merge activity, the final merge, is shown in Fig. 5.5.2. Subfiles are read from each INPUT DEVICE. There is actually only one subfile on each INPUT DEVICE for this example. During the merge, a new ordered file is created and it is placed on one OUTPUT DEVICE. It is clear to the program that the merge is over because there is no output for the other two DEVICES.

### Complex Merges

The balanced merge requires a considerable number of DEVICES. Methods have been devised to reduce the number of DEVICES while maintaining the power of the merge—keeping the order of the merge as high as possible. Most techniques depend on distributing ordered subfiles unequally to the OUTPUT DEVICES concerned.

Although it is beyond the scope of this book to discuss complex merges in detail, we shall pay some attention to the polyphase sort. The full details of its operation along with a discussion of the other complex merges are found elsewhere.†

Figure 5.5.3 indicates the actions which occur during the **polyphase sort** which here uses four DEVICES. We examine this sort with regard to TAPE DEVICES for which it finds the greatest application. In the figure we see that input, the list of unordered records, is provided on the DEVICE T1. We begin examination after the JCL and sort description has been absorbed by the utility that has decided to use the polyphase sort. The internal sort includes distribution. Suppose that the number of records supplied and the size of the buffer areas provided produces exactly thirty-one subfiles of ordered records. These subfiles are distributed to the three output DEVICES, T2, T3, and T4, so that T2 receives thirteen subfiles while T3 receives eleven and T4 receives seven. An algorithm determines how many subfiles each DEVICE receives and in what order.

*first*
*merge*
In the figure we see next how subfiles are merged by a three-way merge. Subfiles are taken one at a time from T2, T3, and T4 and interspersed; the resulting subfile is placed on T1 as shown.

---

† Ivan Flores, *Computer Sorting*, Prentice-Hall, Englewood Cliffs, N.J., 1969.

**Figure 5.5.3.** The polyphase merge

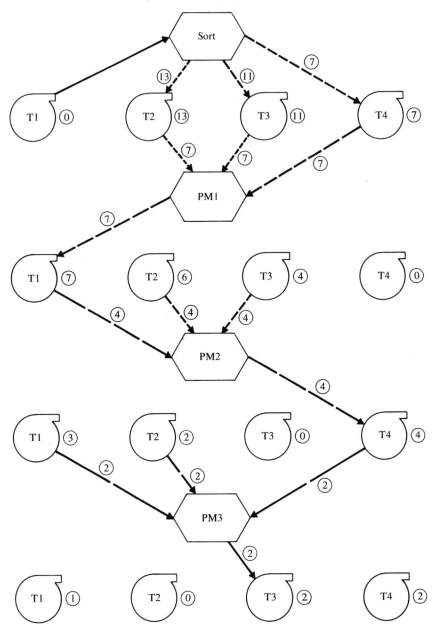

After seven subfiles from each drive are so treated, an end-of-file is noted on T4 since it is exhausted.

*second*          The action at this point is shown in the next frame of Fig.
*merge*           5.5.3. T1 now has seven subfiles, each three times as large
                  as the original ones; T2 has six subfiles left on it; while T3
has four subfiles left on it. Now T4 is empty. It becomes the output unit and T1
now becomes one of the input units. Both T1 and T4 are rewound.

A three-way merge again takes place. This time subfiles from T1, T2, and
T3 are merged. Note that the figure shows the subfiles from T1 as considerably
larger. Merge action continues until T3 is exhausted, thus producing four
ordered subfiles on T4. Each is approximately five times as large as an original
subfile.

*subsequent*      As each merge completes, described as a **stage**, the figure
*merges*          shows the contents of each DRIVE. Thus, at the third stage,
                  T1 contains three subfiles of triple length, T2 contains two
subfiles of unit length, T3 is empty, and T4 contains four subfiles, each five
times as large as the original one. To perform the next merge, the empty
DRIVE becomes the OUTPUT DRIVE and all the others become INPUT DRIVES.
The new OUTPUT DRIVE and the old OUTPUT DRIVE (those that change status)
must both be rewound. Then a three-way merge takes place again.

The remainder of the figure shows the state of each DEVICE at subsequent
stages. At the next to last stage, each input unit contains one subfile. The
last merge creates a single subfile that is entered upon the OUTPUT DEVICE
and signals the completion of the sort.

*advantage*       With four TAPE DRIVES we have been able to provide a
                  three-way merge, thus improving the power and efficiency
of the merge while keeping the number of DEVICES low. Clearly if five DEVICES
were available, we could do a four-way merge, thus obtaining still greater
power while using a number of DEVICES for which we could not even do a
three-way balanced sort.

## 5.6 BINARY SEARCH

### Introduction

Binary search is so efficient because each successive look
cuts in half the number of records to be considered. This quickly homes in
on the desired record. The sole drawback is that calculations are required
between each look, which tends to increase the total search time. Also it

can be used profitably only on a dense ordered file. In all cases, the file must be ordered.

As search begins, we enter the list somewhere near the middle. The *cell* that divides the list in two parts is called the **fence**. By comparing the **fence key** (the key of the record in the fence) with the sought key, we determine whether to look in the sublist above or below the fence.

Take this designated half sublist, upper or lower, and partition it in the same way. This partitioning cell is the **second fence.** It divides *this* half into two quarters and the **second fence key** determines whether the top or bottom quarter is the place to continue the search.

We continue thus, finding eighths, sixteenths, etc., until some fence separates a sublist into a top and bottom sublist, each of which contains *but a single cell.* Then the *next* look will be the *last.* Comparison of the last fence key with the desired key will then either find the desired record or determine that it is absent.

### Illustration

In Fig. 5.6.1 the list **L** is to be examined to find a record with key k. First, find the fence $F_1$; compare its key, $(F_{1K}) = f_{1K}$, to k:

- For $f_{1K} = k$, the record is found.
- For $f_{1K} < k$, the *high half* of **L** becomes $_2$**L**.
- For $f_{1K} > k$, the *low half* of **L** becomes $_2$**L**.

**Figure 5.6.1.** Binary search

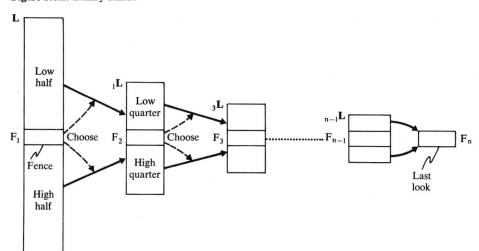

Now find $F_2$ in $_2L$ and compare $f_{2K}$ with k to determine $_3L$ as shown. Then $F_3$ is found in $_3L$ and examined to determine $_4L$. Eventually, find $_{n-1}L$, which consists of three cells with $F_{n-1}$ in the middle. This fence key, if not the one sought, will determine whether the cell above or below will become $F_n$, the last fence. If we are *sure* that the sought record is in L, we don't even have to look at $F_n$. The last look determines whether $(F_n) = f_n$ is the sought record or that it is absent.

As we take more and more looks, it is more and more likely that a fence *has* the sought record—the alternatives to where the record may lie grow smaller and smaller as the search progresses. We may never look at $F_n$ if the sought record is at $F_i$, $i < n$.

### Ideal List Size

The **ideal list size** for binary search is a number that makes the search come out *just right* at each step. To find it, suppose that every search takes just n looks. On the look before the last one the situation is shown in Fig. 5.6.2. The fence cell $F_{n-1}$ divides the sublist $_{n-1}L$ into two parts,

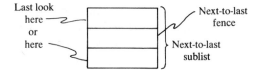

**Figure 5.6.2.** Binary search, next-to-last look

each consisting of a single cell. The key in that cell determines which cell to look at next—and that look must be the last. Thus there are exactly three cells involved.

Just before this, the fence, $F_{n-2}$, divides the sublist, $_{n-2}L$, into two parts, each containing three cells—that sublist contains seven cells as shown in Fig. 5.6.3. In general, $N_I$ (ideal) is given as

$$N_I = 2^n - 1 \qquad (5.6.1)$$

This can be shown at level i as above. To go to the next previous level, $i + 1$, we have a fence with $2^i - 1$ cells on each side, for a total of

$$N_{i+1} = 2(2^i - 1) + 1 = 2^{i+1} - 2 + 1 = 2^{i+1} - 1 \qquad (5.6.2)$$

which is a proof of (5.6.1) by induction.

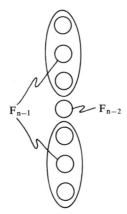

$F_{n-1}$   $F_{n-2}$

**Figure 5.6.3.** Binary search, two
looks before last

### Calculations

For an ideal list size we now examine how a binary search
is performed and the calculations that are required during the search. Let
N be the list size, given by

$$N = num\ \mathbf{L} = 2^n - 1 \qquad (5.6.3)$$

where n is the maximum number of looks required to perform the search
and is given by

$$n = \lceil \log_2 N \rceil \qquad (5.6.4)$$

where $\lceil x \rceil$ means the next integer after x for x not an integer or x otherwise.
The ordinal number of the cell that separates the list into two exactly equal
parts $N_1$ is give as

$$N_1 = 2^{n-1} \qquad (5.6.5)$$

Thus $N_1$ is the number of the cell within the list that becomes the first fence, or

$$F_1 = L_{N_1} \qquad (5.6.6)$$

Figure 5.6.4 illustrates how the first fence $F_1$ separates the list into two
equal parts. The fence creates a low and a high sublist, each of equal size as
shown in the figure. The next action is determined by examining the key of
the record found in the fence. If the key is the same as the key sought, we
have found the desired record. Otherwise we shall look in one of the sublists
created by the partitioning introduced by the fence itself. If the key sought

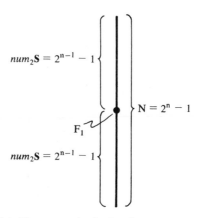

**Figure 5.6.4.** Binary search, the first fence creates two halves.

is smaller than the key of the fence record, we shall use the lower sublist next:

$$k < (F_{1K}): \quad _2S = \{L_1, L_2, \ldots, L_{N_1-1}\} \qquad (5.6.7)$$

If the key sought is higher than the key of the fence record, we shall look in the higher sublist, thus:

$$k > (F_{1K}): \quad _2S = \{L_{N_1+1}, \ldots, L_{N-1}, L_N\} \qquad (5.6.8)$$

**second fence**   There are two alternatives for the second fence, as noted in (5.6.7) and (5.6.8), according to which sublist we are using.

Again because our list is of the ideal size, it is easy to find a fence that separates the second sublist into two sublists of exactly the same size, thus:

$$N_2 = 2^{n-2} \quad \text{or} \quad N_2 = 2^{n-1} + 2^{n-2} \qquad (5.6.9)$$

where $N_2$ is the cell number of the second fence.

To summarize this in a single equation, we have

$$N_2 = N_1 \pm 2^{n-2} \qquad (5.6.10)$$

where we choose the positive or negative sign according to whether the low or high sublist is used.

**general fence**   Suppose we have just been working with the $i - $ 1st fence, namely $N_{i-1}$, and we wish to find the next fence, $N_i$. We simply add or subtract the proper constant, thus:

$$N_i = N_{i-1} \pm 2^{n-i} \qquad (5.6.11)$$

Note that $N_i$ is the *number* of the cell. It is used as a subscript when designating the fence, thus:

$$F_i = L_{N_i} \qquad (5.6.12)$$

**looks**    The maximum number of looks is n. As we proceed further and further along the way, however, the probability that any look is successful increases greatly. This is because more and more of the list has been removed from consideration, it being known that the desired record could not be contained in that portion of the list. It has been shown mathematically that the expected number of looks is $n - 1$:

$$L \doteq n - 1 \qquad (5.6.13)$$

**example**    Figure 5.6.5 shows a list **L** of fifteen records, an ideal size. Let us search this list for ilze. Applying (5.6.6) we find that the eighth record is the fence record, and so we examine hank. We note that ilze is greater than hank and so we choose the top sublist. We use the plus sign in (5.6.10) to examine max. Since ilze is less than max, we look at the lower sublist for the next fence. Using the minus sign in (5.6.11) we come to jill. Since ilze is less than jill, we use the lower sublist again. Since that sublist consists of only one record, that record is the fence and, as we see in the figure, turns out to be the sought record ilze.

It is easy to see from Fig. 5.6.5 that a search for jill will turn her up on the third look. Of course, hank will turn up on the first look.

**Figure 5.6.5.** Binary search for ilze

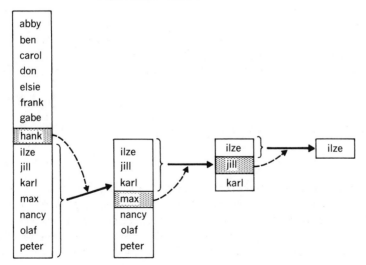

A search for an absent record requires all n looks. A search for dave is shown in an illustration taking a different form in Fig. 5.6.6. Here we see that we look in order at hank, don, ben, and carol. Since we can go no further and carol is not the record sought, we can only conclude that dave is absent.

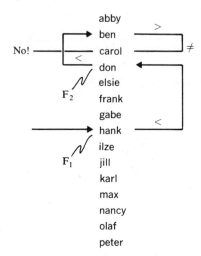

**Figure 5.6.6.** Binary search for dave who is absent.

## 5.7 BINARY SEARCH—
## ANY NUMBER CAN PLAY

### Problem

We have examined binary search when N, the size of the list, was ideal. Generally, the list to be searched is not an ideal size. When this is the case, the number of looks n is still governed by the inequality

$$2^{n-1} < N < 2^n \tag{5.7.1}$$

or, stated another way, we have

$$n = \lceil \log_2 N \rceil \tag{5.7.2}$$

The principle of binary search is the same regardless of the size of the list. We divide the list into two "halves" of comparable size that differ from each other by no more than 1 and choose a half list by comparing the key sought with the fence key. When we deal with an ideal number, the fence divides the list into equal sublists, each also of ideal size. This is no longer true, and the fence that we choose will necessarily produce sublists of unequal size.

There are three principal ways described here for finding fences and sublists. Described briefly, these are as follows:

- Use upper and lower limit cells to define each.
- Keep track of and manipulate the current size of a sublist.
- Modify the ideal number method.

### Sublist Limit Method

In this method we use two markers or registers at each level. Let us call the original list $L$. $F_1$ is the fence for $L$ and it divides $L$ into two parts: a lower sublist, $_LL$, and a higher sublist, $_HL$. One of these will be selected by examining the fence key. Let us call the one selected $_1L$. The ith sublist is bounded by an upper and lower cell, and we have the following:

- $N_{Li}$ is the number of the lowest cell in the ith selected sublist.
- $N_{Hi}$ is the number of the highest cell in the ith selected sublist.

After defining a sublist by its boundaries thus, average these boundaries to come up with a fence that approximately divides the sublist into two equal parts. Thus we have

$$N_{i+1} = \lfloor (N_{Li} + N_{Hi})/2 \rfloor \qquad (5.7.3)$$

where $\lfloor x \rfloor$ means the integer just less than x or x itself if x is an integer.

The *ordinal number* of the fence with regard to the *original list* acts as a subscript. The cell address is a function of the starting point of the list and the size of the record in primary units (such as bytes). Putting this all together, we have

$$F_i = L_{N_i} = L + C(N_i - 1) \qquad (5.7.4)$$

where C is record size in bytes.

*set up*　　　　To set up, we initialize markers L and U to the beginning and end of the original list, thus:

$$1 \rightarrow N_L; \quad num\ L \rightarrow N_H \qquad 1 \rightarrow L; \quad num\ L \rightarrow U \qquad (5.7.5)$$

As search progresses, compare the key sought with the key contained in the fence for this level. Use F to hold the fence number calculated as in (5.7.3) thus:

$$\lfloor [(L) + (U)]/2 \rfloor \rightarrow F \qquad (5.7.6)$$

At level i, suppose that the key sought is *less than* the fence key. Choose the lower sublist. The lower marker for this sublist will then be the same as the lower marker for the new sublist, however, the higher marker will point to the cell just below the fence. Don't alter L; just change U, thus:

$$k < (F_{iK}): \quad (F) - 1 \rightarrow U \qquad (5.7.7)$$

If the key sought is greater than the fence key, the new sublist will *begin* right after the fence. It will end at the top of the current sublist, so that we only change L:

$$k > (F_{iK}): \quad (F) + 1 \rightarrow L \qquad (5.7.8)$$

This is illustrated in Fig. 5.7.1, where the boundaries of the current sublist appear on the left, along with the pointer to the fence, and the boundaries for the new sublists appear on the right.

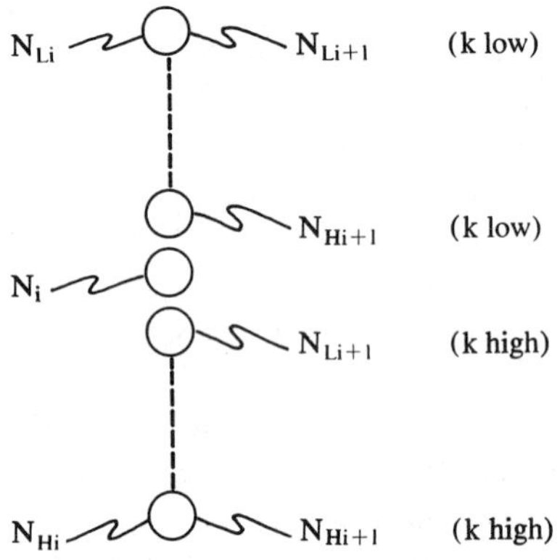

**Figure 5.7.1.** Sublist limit method shown symbolically

*example*    In Fig. 5.7.2 we find a list of twenty records. We shall search for pierre. The upper and lower numbers of the original list are one and twenty. Applying (5.7.6) we find that the fence item has the number 10 and hence we examine jill. Since pierre is greater than jill, we reset the lower boundary to eleven. Applying (5.7.6) again, we arrive at olaf (15). pierre is still higher, so again we reset the lower boundary, using (5.7.7), to sixteen. Calculation of the fence takes us to ralph at 18, who is greater than pierre. Now we readjust the upper number to seventeen. With a lower number of sixteen and an upper number of seventeen, the next fence calculation takes us to pierre!

### Sublist Size

Instead of keeping track of the upper and lower limits of each sublist, this method keeps track of the fence and the size of each sublist.

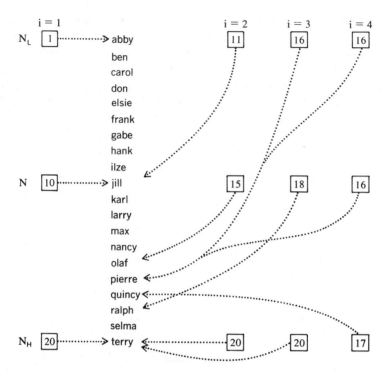

**Figure 5.7.2.** Sublist limit method, an example

*set up*           As the search begins, the initial sublist size $S_1$ is actually the list size and the fence is placed approximately halfway down the list, so that we have

$$N_1 = \lceil N/2 \rceil; \quad S_1 = N \tag{5.7.9}$$

*subsequent*      Given the size of the list, the size of the first half list is
*sublists*         indicated by $S_2$ and is the size of the list divided by 2 and rounded down. Thus we have the following equation:

$$S_2 = \lfloor N/2 \rfloor = \lfloor S_1/2 \rfloor \tag{5.7.10}$$

Given the size $S_i$ of the ith sublist we calculate the size of the next sublist, $S_{i+1}$ by dividing by 2 and rounding down, thus:

$$S_{i+1} = \lfloor S_i/2 \rfloor \tag{5.7.11}$$

*fence*           The determination of a fence is made by using the current fence and proceeding forward or backward from this fence by the size of the *next* sublist. Notice that in the formulas sometimes we round up while at other times we round down. This is because we want to space across the cells to a fence so that the sublist remaining is one smaller than the interval required to get from the current fence to the next fence. This is illustrated in Fig. 5.7.3. Thus the fence whose number is $N_i$ divides the sublist of length $S_i$ into two portions, each of length $S_{i+1}$. The distance of the next fence, whose number is $N_{i+1}$, from this fence, numbered $N_i$, is half of $S_{i+1}$ rounded up, as shown in the figure or as in

$$N_{i+1} = N_i \pm \lceil S_{i+1}/2 \rceil \qquad (5.7.12)$$

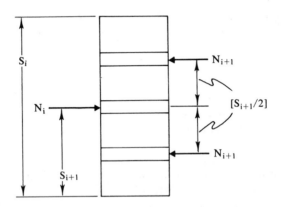

**Figure 5.7.3.** The fence determined by sublist size

*choice*    We choose the next fence as before, according to whether the key sought is less or greater than the fence key. Then the next fence number is given by

$$k < (F_{iK}): \quad N_{i+1} = N_i - \lceil S_{i+1}/2 \rceil \qquad (5.7.13)$$

and

$$k > (F_{iK}): \quad N_{i+1} = N_i + \lceil S_{i+1}/2 \rceil \qquad (5.7.14)$$

*termination*    The size of the last sublist will be 1. If we take half of this and round down, we will find 0, so that

$$S_n = 1; \quad S_{n+1} = 0 \qquad (5.7.15)$$

*example*    In Fig. 5.7.4 we see the same list as appeared in Fig. 5.7.2, which we shall now search by the sublist size method. Fence numbers and sublist sizes are indicated in the figure in the squares, with pointers showing how they distinguish the fences and sublists to which they pertain. As before, our search is for pierre.

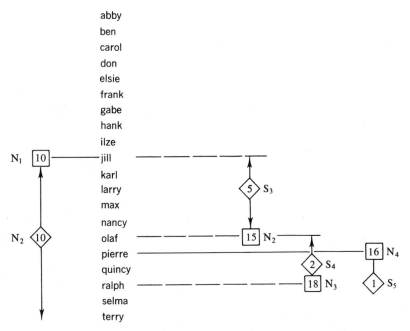

**Figure 5.7.4.** Sublist size search example

### Fence Number Always a Power of 2

For this technique we pretend that we have a list whose size is ideal (when actually we don't). The number of looks n is defined by (5.7.2) as before, where N is given by (5.7.1). We pretend that we have the full list size and set up our fence where the middle would be for the ideal list. Actually, this divides the list into two parts that may be widely different in size. The number $N_1$ of this first fence is then given as

$$N_1 = 2^{n-1}, \qquad 2^{n-1} \le N < 2^n \tag{5.7.16}$$

*lower*
*sublist*

After finding the first fence, $F_1$, compare the key sought with the fence key, as always. Should the key sought be smaller, the lower sublist is selected. This sublist is of ideal size. Hence the next fence will be determined by

$$k < (F_{iK}): \quad N_2 = 2^{n-2} \tag{5.7.17}$$

Further, subsequent fences will be determined as in the previous section by simply adding or subtracting the next lower power of 2. Then the ith fence is determined thus:

$$k < (F_{iK}): \quad N_i = N_{i-1} \pm 2^{n-i} \tag{5.7.18}$$

where $+$ or $-$ is selected by comparison of the key sought with $(F_{i-1K})$.

***high***　　　　The first fence has created a second or high sublist, which
***sublist***　　　we shall use should the key sought be higher than the fence
　　　　　　　key. The number of records in this sublist is probably much
smaller than in the lower sublist. We now search for the highest power of
2 to add to the fence number, to create a new fence that still lies within the
sublist. Suppose the number of records in the ith high sublist is given by
$S_{iH}$ where

$$2^q \leq S_{iH} < 2^{q+1} \tag{5.7.19}$$

In other words, $2^q$ is smaller than or equal to this number, but $2^{q+1}$ is greater.
Then we find the next fence by adding the qth power of 2 to the current fence
whose number is $N_i$, thus:

$$N_{i+1} = N_i + 2^q \tag{5.7.20}$$

This is shown in Fig. 5.7.5.

Once we have started using the higher sublist, a similar situation prevails
as when we began:

- The fence divides this sublist into two parts.
- The lower part contains an ideal number of cells.

**Figure 5.7.5.** Power of two method

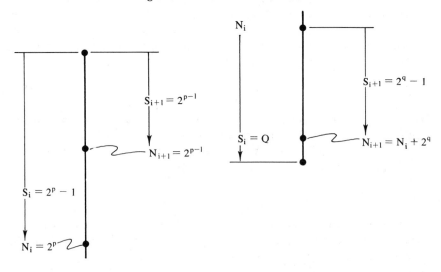

- The upper sublist contains a number of cells that at the moment is not known but is probably not ideal.

In general, when we choose the lower sublist, the situation becomes ideal and we follow the method of Section 5.6. When we choose the higher sublist, we deal with a nonideal number and use the method of (5.7.19) and (5.7.20).

*example*    In Fig. 5.7.6 we see the list used in previous examples containing twenty records. It will be very easy if we look for pierre because the first fence selected will contain him, so why not look for quincy?

As shown in the figure, for the first fence, $N_1$ is 16, and the fence contains pierre who then divides the list into two parts. The lower part is of ideal size, containing 15 cells. The higher sublist contains four cells. Since quincy is greater than pierre, we select the upper sublist.

Now the highest exponent of 2 that we can use is 2 since $2^2 = 4$, the size of the upper sublist. $N_2$ is hence 20 and we next look at terry. This time we choose the lower sublist since quincy is smaller than terry. This new sublist is of ideal size, 3. Hence succeeding fences will be chosen by the procedure of Section 5.6. $N_3$ takes us to ralph and $N_4$ takes us to quincy. This is our last look, so we are happy to find that quincy occupies $L_{17}$.

**Figure 5.7.6.** Power of two example

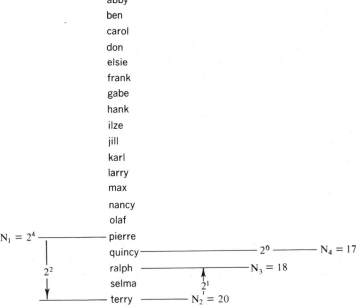

**PROBLEMS**

5.1   For the semidense list:
(a) What need does it fulfill?
(b) Characterize the two parts.
(c) How is the boundary recognized?

5.2   To append record r to the semidense ordered list $L$ we find the target $L_T$ that divides $L_D$ into two parts, $L_S$ with records smaller than r and $L_G$ with records greater than r, neither of which includes $L_T$.
(a) How is $L_T$ found originally?
(b) Describe symbolically the sublist containing the records that must be moved up one cell, $L_U$.
(c) Why must the record from the *last* cell in $L_U$ be moved into $L_E$ first?

5.3   Suppose there is another set of empty cells in $L$ that *precedes* $L_D$; call it $L_E'$. Now append r by moving back $L_B$ so as to free $L_T$.
(a) How do we define $L_B$?
(b) Which record of $L_B$ is moved first into $L_E'$?

5.4   Consider the list, $L$, consisting of $L_E'$, $L_D$, and $L_E$ as described above and in Prob. 5.2 and 5.3. To append r again, find $L_T$. Now move $L_B$ back or $L_U$ up according to which requires the least movement. How is this decided?

5.5   To *delete* record r from $L$:
(a) How do we find $L_T$?
(b) How do we remove r from $L$ consisting of $L_D$ and $L_E$?
(c) How do we remove r from $L$ consisting of $L_E'$ and $L_D$?
(d) How do we remove r efficiently from $L$ consisting of $L_E'$, $L_D$, and $L_E$?

5.6   For the double ended semidense list of Fig. 5.1., show the result of each of the following operations in this order by making five separate sketches:
(a) Delete ben.
(b) Add ken.
(c) Add bernice.
(d) Add dan.
(e) Delete larry.

5.7   The first record in $I$ is frank; the next is carol; then comes joyce, howie, ben, gabe, olga, larry, nancy, ann, errol, karl, irv, dave, and marv in that order. Show in fourteen separate frames that the resulting list $L$ as a sort by insertion occurs.

| Address | Contents |
|---------|----------|
| 79 | @ |
| 80 | @ |
| 81 | ann |
| 82 | ben |
| 83 | carol |
| 84 | dave |
| 85 | errol |
| 86 | frank |
| 87 | gabe |
| 88 | howie |
| 89 | irv |
| 90 | joyce |
| 91 | karl |
| 92 | larry |
| 93 | marv |
| 94 | nancy |
| 95 | olga |
| 96 | @ |
| 97 | @ |
| 98 | @ |

**Figure 5.1.**

5.8   For the list I in Fig. 5.2., show in fourteen frames how a sort by selection and exchange is done.

5.9   For the list I in Fig. 5.3., show how a sort by selection with replacement is done where new records com from L in this order: errol, karl, irv, dave, marv, amy, fred, clara, vince, jim, abe, inge, faye, mary, albert, sandy, herb, mel. Show S and I after the first run is produced.

**Figure 5.2.**

| Address | Contents |
|---------|----------|
| 107 | frank |
| 108 | carol |
| 109 | joyce |
| 110 | howie |
| 111 | ben |
| 112 | gabe |
| 113 | olga |
| 114 | larry |
| 115 | nancy |
| 116 | ann |
| 117 | errol |
| 118 | karl |
| 119 | irv |
| 120 | dave |
| 121 | marv |

| Address | Contents |
|---------|----------|
| 218 | frank |
| 219 | carol |
| 220 | joyce |
| 221 | howie |
| 222 | ben |
| 223 | gabe |
| 224 | olga |
| 225 | larry |
| 226 | nancy |
| 227 | ann |

**Figure 5.3.**

5.10 Revise the rules for two-way merging so that twinning (records with the same key) may occur.

5.11 Show the calculation and the cells visited for a binary search of the list in Fig. 5.1. to find
   (a) ben,
   (b) arlene.
   (c) marv,
   (d) steve.

5.12 For the list of Fig. 5.4., use the sublist limit method to find
   (a) ben,

**Figure 5.4.**

| Address | Contents |
|---------|----------|
| 604 | ann |
| 605 | ben |
| 606 | carol |
| 607 | dave |
| 608 | errol |
| 609 | frank |
| 610 | gabe |
| 611 | howie |
| 612 | irv |
| 613 | joyce |
| 614 | karl |
| 615 | larry |
| 616 | marv |
| 617 | nancy |
| 618 | olga |
| 619 | pat |
| 620 | quincy |
| 621 | roger |
| 622 | sally |
| 623 | tom |
| 624 | ursala |
| 625 | vic |
| 626 | walter |

    (b) arlene,
    (c) sally,
    (d) steve,
    (e) vic.

5.13 Repeat Problem 5.12 using the sublist size method.

5.14 Repeat Problem 5.12, this time using the power of 2 method.

CHAPTER  6

# Linked Lists

## 6.1 INTRODUCTION

### Sequential Files

*problems*    In some cases we are as concerned with file fabrication and
              alteration as with reference or posting (individual record
alteration). *When file alteration predominates, linked lists can provide an
advantage over the sequential list.* We have examined dense ordered lists. We
have found effective methods for searching them, such as binary search that
cuts down the number of looks needed to find the desired record. Having
found a record, it is easy to alter it, provided the new information fits into the
record. It is when insertion and deletion (file alteration) are required that we
get into trouble with the ordered list.

*insertion*    To append a new record to an ordered file, we first search
              the file to find the place where the new record belongs.
Since the file is dense, there is obviously no free room for the new record.
To create a place for the record, move all the records one cell ahead in the
list. Then move the record in the chosen cell to the next cell. Place the new
record into the "hole" thus created.

Sometimes an improvement can be made when the record is to be placed
near the beginning of the file. It may be possible to move up all the records
above it one place higher. In any case, as you can see, we must do a lot of
information movement to place a new record into the file.

156

*deletion* To remove a record from the file, again search for the record. Once found, delete the record by "crossing it out"—inserting a symbol such as ¢ in the key that tells us that there is no longer information at this place in the file.

After many deletions the dense file is really no longer dense. It has many holes, wasted space left over from expired records. Eventually the whole file will have to be rewritten to get rid of the holes.

*complex* An ordered list without special artifices can only store a
*structure* linear file. A tree or a transitive graph structure cannot be simply represented using a linear ordered list. The linked list structure plus the simple devices discussed in Section 6.6 convey and store in MEMORY the tree and the transitive graph file.

### List Area

In Fig. 6.1.1, we examine a linked list as it appears in MAIN MEMORY. There is an area assigned to the list called the **list area**, chopped up into units of record size, each called **cells**. Cells of linked lists are shown as rectangles with a triangle affixed to one narrow end.

Notice that the linked list in Fig. 6.1.1 is a loose list. We defined a cell as a set of contiguous words or bytes of record size.

There is another cell, the **head**, outside the list that points to the first cell of the linked list. The head need not be so large as a list cell since it only contains a pointer to the cell with the first record of the file. Each cell is pointed to by its predecessor with respect to the file. Of course we must have some way of distinguishing the beginning and end of the file; this is discussed later.

### The Record

Figure 6.1.2 pictorially conveys a record a at location A in a linked list **A**. The record has many fields, each of which may be distinguished by a subscript. This subscript is a digit for data fields. Thus the data fields are $a_1$, $a_2$, $a_3$, etc.

Two other fields are of more importance to us. The key field, $a_K$, permits us to arrange the records in the linked list as they should appear in the file; the pointer field, $a_P$, tells us the address of the next record. It is the link. We also refer to these fields in the cell by the names $A_K$ and $A_P$, respectively. To simplify our explanation of the linked list, we combine all the data fields into one symbol as discussed below.

The linked list may not store any data but rather a data pointer to a fixed

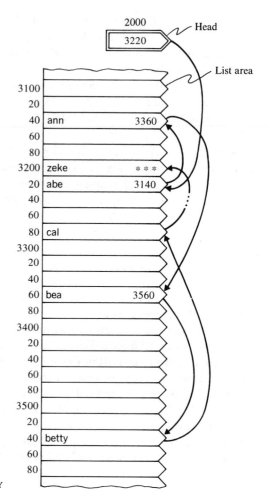

**Figure 6.1.1.** A linked list in MEMORY

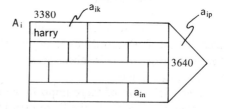

**Figure 6.1.2.** A record a at location **A**

area in MEMORY where the record is located, as in Fig. 6.1.3. Manipulation of the linked list is simplified because the actual datum stays fixed. All work is done with the smaller pointer entry for this linked directory file structure, examined in Section 8.3.

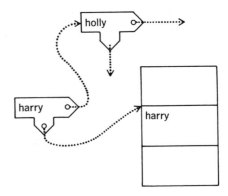

**Figure 6.1.3.** Cell with a data pointer

### Record Surrogate

Figure 6.1.4 shows a record surrogate, a simplified diagram that is sufficient to convey the linked list structure without the additional complication of considering the entire record. It is also a neat method of handling variable length records using the data pointer as mentioned.

**Figure 6.1.4.** A record surrogate

```
        3380
   harry        3640
```

The reader should keep prominent in his mind that although we deal in what follows with this record surrogate, whatever is said applies equally well to an entire record of fixed size.

In Figure 6.1.5 we see a head **f** which points to a cell, A, which contains a datum (key), a, and also a pointer, B. The next record in the file is located at B. It contains a datum, b, and a pointer, C, to the next record at C.

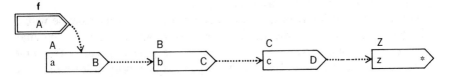

**Figure 6.1.5.** Diagram of a linked list

### Definition

A **linked list** contains a file, often ordered, stored in MAIN MEMORY so that the records in the list are not generally in contiguous locations but each contains a pointer to the next record according to the order relation (if any) on their key field(s).

The reader should distinguish between the physical list itself and its contents. The physical list contains cells and is ordered only by virtue of the location of these cells and the addresses hence assigned to them. The contents of the list assume the ordering imposed by the pointer chain, but this ordering need not be inherent in the data itself. It is only a ranked linked list what has content ordering. It is possible to have a linked list where no order is imposed by the data itself. The list may be created according to arrival and used in that order (first in, first out; FIFO) or in reverse order (last in, first out; LIFO).

## 6.2 USING THE LINKED LIST

### Merits

The linked list is especially advantageous for file maintenance. We shall see in this dection how easy it is to append new records and to delete expired records. It is also easy to search a linked list sequentially. When we reach a cell, we simply extract the pointer to the next cell; pointer extraction takes only a command or two.

The disadvantage of the linked list is that we cannot employ sophisticated techniques such as the binary search for quickly finding desired records.

### The List

Every list begins with a head, a cell usually external to the list area, which points to the beginning of the list or else indicates that the list is empty. The list consists of cells, each containing at least a key, the link, and either a data pointer field or data fields.

In figures that follow we use a characteristic cell shape—a rectangle with a triangle grafted to one of the narrow sides—as shown in Fig. 6.2.1. We may use some symbolic name to address the list and generally this is the same name that is applied to the head. In the figure, keys appear in lowercase san serif type to distinguish them from the locations of cells in capitals. Often the cells are given names that correspond to the key of the record they contain. Thus art is the key of the record contained in ART.

For the sake of simplicity our linked lists show only keys and pointers. The symbolic pointer name is naturally found at the pointer end of the cell in the figure. Thus at ART we find a record with key art and a pointer BEA because the next cell in the list is at BEA and has the key bea.

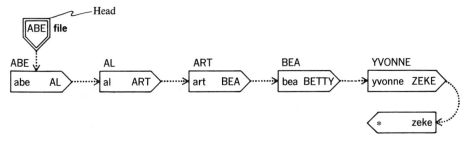

**Figure 6.2.1.** A diagram of a file

***termination***    It is important to know when we have reached the end of
the file; the file in Fig. 6.2.1 ends at the cell ZEKE; it contains a record with the key zeke and a terminal pointer. We use * to indicate
this terminal pointer. When * is detected, it signals the end of the list.

### Appending

In Fig. 6.2.2 we see a file called **file**. Six of the records of this
file are shown. bea points to betty. In between these two we would like to
insert another record with key belle. When we receive our instructions, we also
receive the record key belle.

***space***    To add records, there exist available empty cells in the list
area. Suppose for the moment that we have such a set of
cells and that they are contiguous. Further, suppose that they have been

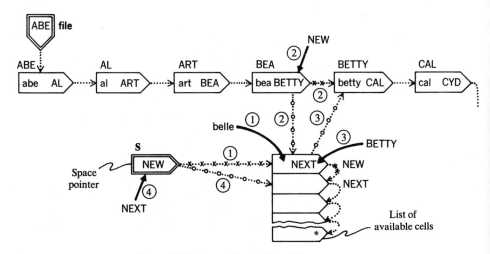

**Figure 6.2.2.** Appending belle to **file**

initialized so that each points to the next and that there is a pointer, a head called **S**, that points to the next available cell in this empty cell list, as shown in Fig. 6.2.2.

***action***    To append belle to our file, first place the whole record belle into the cell pointed to by **S** (1). Then determine that belle follows bea. Next enter the pointer to this new cell that we shall call NEW into the pointer part of BEA (2); the pointer previously contained BETTY. We have now pointed BEA to NEW.

Next we must point NEW to BETTY. We picked up the pointer BETTY and now we place it into the pointer portion of NEW (3).

Finally we update the space list. NEW used to contain a pointer NEXT. This (NEXT) now goes into **S** to point to the next available cell.

Figure 6.2.2 shows unaltered pointers as dotted lines; new pointers are dots mixed with o's; expired pointer lines consist of dots and x's.

**Deletion**

To delete art, first find ART, which contains art (1). Then proceed as in Fig. 6.2.3 by extracting the pointer in ART, which is BEA (2). Then go to the predecessor of ART, which is AL. Note that the pointer in AL is ART (3). Change this pointer by inserting BEA to replace ART (4). Now AL points to BEA (5). Notice that ART is detached and hanging. He still points to BEA. This does no harm; he is no longer part of the linked list. We shall see means for coping with this detached cell later.

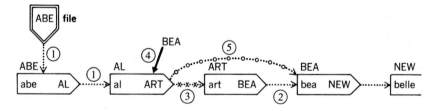

**Figure 6.2.3.** Deleting art from file.

**Search**

Search a file by traveling from one record to the next using the pointers. To present this symbolically, use the operator *point* to extract a pointer from its cell. To start a search, extract the pointer from the head, thus:

$$point \textbf{ file} = ABE = [abe] \tag{6.2.1}$$

To find the next record, extract the pointer from the cell just found, thus:

$$point \ ABE = AL \tag{6.2.2}$$

Another way to designate the location of the cell holding the second record is

$$point^2 \textbf{ file} = AL \tag{6.2.3}$$

A search examines the key of successive records, comparing them with the sought key.

**equal**         To perform a *search equal* for a particular key we must find successive records in the file and compare their keys with the one furnished. The operator *esearch* is also used to search a linked list for equal (as well as the sequential list). To the left of the operator is the key upon which the search is made; to the right of the operator is the file being searched or the head of that file. We indicate a search for a record with the key cal thus:

$$cal \ esearch \textbf{ file} = CAL \tag{6.2.4}$$

Notice that the operator turns up the location CAL, which contains the record.

When the record sought is absent from the file, the operator turns up no cell but the terminal pointer, *, instead, thus:

$$bob \ esearch \textbf{ file} = * \tag{6.2.5}$$

**equal**
**greater**        To append a new item to an *ordered* linked list file, search the file using the key of the new item but this time looking for a record with a key just greater than the one furnished. We use the *egsearch* operator to *search equal greater*. Thus if we provide the key bob and use *egsearch* on the file, it will turn up the cell containing a record before which the new record should be inserted, thus:

$$bob \ egsearch \textbf{ file} = CAL \tag{6.2.6}$$

*file length*     Consider a file that contains n records. As we apply the *point* operator, we obtain valid pointers for successive records of the file; however, the last record contains the terminal pointer, *. Thus we have

$$point^n \textbf{ file} = * \tag{6.2.7}$$

## 6.3 THE LIST AREA

The programmer establishes the area for a file to be kept as a linked list. He then places the file as it now stands into this area. Thereafter he is able to modify the file by techniques discussed earlier. Maintenance of the list area depends on the means used for appending and deleting. There are two techniques for doing this:

- The space pointer technique requires a single external pointer cell to keep track of available contiguous space.
- The space list is a separate linked list for keeping track of all empty cells and available space (see Section 6.4).

### The Space Pointer Technique

For this technique at the outset we set outside a linked list area **L** of sufficient size to accommodate the file and allow for anticipated growth. The area is now empty or contains garbage. Assume also a file, **f** elsewhere that is in key order and needs to be placed into the list area, **L**. Establish a head which symbolically bears the name of the file, **f**, and which will point to the beginning of the file in the list area, **L**. Now enter the file into the list.

List initialization is best done concurrently and sequentially. The records are placed in their present order into sequential cells in the list area. As each record is placed into a cell, a pointer is established from that cell to the next (and in this case adjacent) cell. This continues until the last record of the file is placed into a cell as shown in Fig. 6.3.1 along with a terminal pointer so that searches will end there.

Two further actions are required:

- The head, **f**, is set to point to the first record (and cell) in the file.
- The space pointer, **S**, is set to point to the first available cell in the list area (the one after the last cell used).

The graph of the linked list immediately after initialization is presented in Fig. 6.3.2:

- Solid arrows show that cells are adjacent.
- Dotted arrows show that records are linked together by their pointers.
- The remaining empty cells are pointed to by the space pointer, **S**, and are noted as empty by the solid circles.

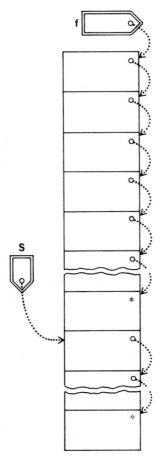

**Figure 6.3.1.** The file **f** has just been put into **L**.

- Available cells are associated only by adjacency as shown by the arrows from one to the next in the figure.

**Appending and Deleting**

Let us see how to append and delete records using the space pointer technique.

*delete*     In Fig. 6.3.3 we have deleted record $f_2$. Formerly, $f_2$ was preceded by $f_1$ and followed by $f_3$. To perform the deletion, the pointer in $f_1$ is altered so as to point to $f_3$. Nothing has to be done to $f_2$ itself; it is unreachable as far as the user of the file **f** is concerned.

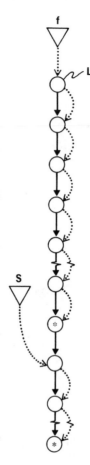

**Figure 6.3.2.** Graph form of the new
file, **f**

***append***        Figure 6.3.3 also shows the file f, after the record f′₃ has
been appended between f₃ and f₄. Here are the actions to be
performed to obtain this effect:

- Search the file to find that f′₃ belongs after f₃ but before f₄.
- Go to the space pointer and find the necessary available cell.
- Note the location of the new cell and advance the space pointer to point
  to the *next* available cell.
- Put the new record into the available cell.
- Point f₃ to the new cell containing f′₃.
- Point f′₃ to f₄.
- f₄ is not affected in any way.

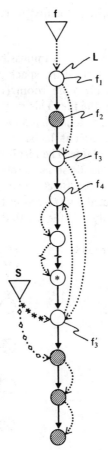

**Figure 6.3.3.** The file **f** after $f_2$ has been deleted and $f_3'$ has been appended.

### Loose List

During the maintenance of a linked list file, deleting and appending will occur:

- Deleting creates **holes**—cells that do not contain useful information but are unavailable for further use.
- Appending uses up cells from the available area in the list area.

As this occurs, the space pointer is moved further and further down in the list area. Eventually, there will come a time when all the available cells are used up. There are still many holes in the list but these are not available. The file can no longer be properly updated when the situation in Fig. 6.3.4 has been reached.

***recovery***     In Fig. 6.3.4 it is impossible now to append new records to our file since the space pointer now contains a terminal character indicating that no more room is available. There is no way to recover except by copying the file. This may be done in the existing list area or the file may be copied into a new list area. The result is to **compact** the file, creating another available area at the bottom of the list obtained by releasing the holes previously unavailable. The result appears like Fig. 6.3.1.

The action of copying the file so as to release the holes, the cells that were previously unavailable, is sometimes called **garbage collection**. The garbage or holes seem to be collected and placed at the bottom of the list where they are now available—actually it is the file that has been collected or rewritten.

**Figure 6.3.4.** There is no more room in L for new records of **f**.

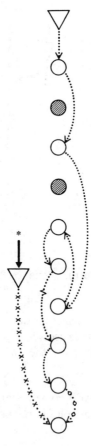

### Variable Record Length

The linked list technique can handle variable length records. How then shall we handle available space? The space pointer technique is adaptable to variable record lengths. When the list area is initialized, the ordered variable length records can be entered into the list area in the same way described for appending new records below.

To append a new record, the space pointer is found and the desired quantity of space is withdrawn. It is a simple matter to update the space pointer variably by the number of bytes withdrawn. Of course, deletion now produces holes of variable size. These holes are not reusable. They are not a problem until the list becomes full. This is coped with using garbage collection as discussed earlier.

The space list concept described in Section 6.5 requires a preallocation of cells and hence implies fixed cell size and thus fixed record size. Of course, the fixed cell could hold a variable size record smaller than the cell but this would hardly provide an economy.

## 6.4 FIXED FORMAT LISTS

### Accessibility Limitations

**Fixed format lists** have accessibility limitations—only certain cells in the list may be accessed. These limitations may stem from two sources:

- Physical constraints—the medium that stores the list has an inherent access constraint such as destructive readout.
- Need and efficiency—lists of certain types only require limited accessibility. A list structure can be more efficient if we do not furnish additional accessibility.

*names*     There are two kinds of fixed format lists that we discuss here:

- **Queues** provide opposite end accessibility: input at one end, output from the other.
- **Stacks** provide single end accessiblity: records can be entered or withdrawn from only one end.

A fixed format list works best when restrictions imposed on its use and updating suit the application. The benefit is that less hardware and software

are needed. The major restriction can be simply stated as "use implies update." To restate this, we have the following:

1. Accessing a record in a fixed format list deletes this record.
2. Search is not possible—the only accessible record(s) is that at the list output.
3. Appending means input—the only way that a record may be entered onto the list is at the input.

The fixed list structure could be modified so that appending and deleting is done by rules 1 and 3 above—search and accessing are possible for interior items in our list (rule 2 would not apply). To reiterate, however, interior items are not accessible in any fixed format lists used here.

### Implementation

The fixed format list is best implemented by a linked list. It is also an adjunct for accessing the linked list. Hence it is appropriate to discuss it here.

### Queue

A **queue** is a list with both ends accessible as shown in Fig. 6.4.1. In that figure, the queue is named **Q**. Notice that it has *two* pointers: the **front pointer** points to the front of the list; the **back pointer** points to the end of the list. The records are entered at the back (right) and withdrawn from the front (left). *Two* pointers are necessary since both ends of the list are used. The pointer *is* the head.

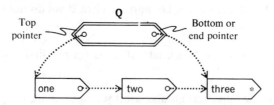

**Figure 6.4.1.** A queue called **Q**.

The queue is used when records must be serviced in order of receipt as, for example,

- a list of areas (buffers) to receive records from a serial file;
- a list of areas to be written onto a serial file.

If the order were not maintained, the information would not be used properly.

There are pointers between records in the list as shown in the figure. These pointers might only be physical adjacency of the records on the medium or they might be actual pointers when a queue is implemented using linked lists.

***insert***        Figure 6.4.2 shows how record FOUR is inserted into our queue. It is placed after the most recent item in the queue. Since THREE is the most recent item and FOUR is being entered, FOUR becomes the item now pointed to by **back pointer Q**. This can be stated symbolically as

$$\text{FOUR} \rightarrow \text{Q} \tag{6.4.1}$$

We enqueue FOUR onto **Q**.

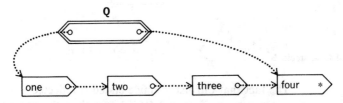

**Figure 6.4.2. Q** after FOUR has been added (*in*ed).

***access***        Figure 6.4.3 shows how the next record from **Q** is found and (necessarily) removed from the list. We state this in our notation as follows:

$$\text{Q} \rightarrow \text{ONE} \tag{6.4.2}$$

This states that ONE has been accessed from the queue **Q**. We dequeue **Q**.

There are other names for queues, the most important of which is **FIFO list**. These initials stand for *first in, first out*. They convey that the oldest record, the one that was entered first, is the one to leave first.

**Figure 6.4.3. Q** after ONE has been removed (*out*ed).

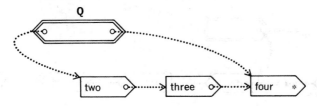

### Stack

A **stack** is a *last in, first out* or **LIFO list**. This conveys that the most recent item is the first one to be withdrawn. A name that I prefer and shall use frequently is **push-down list**:

- **push** is to enter a record into a push-down list.
- **pop** is to access and delete a record from a pushdown list.

Imagine the plate dispenser in a cafeteria. When we take a plate off the top, the stack pops up another plate to meet the next user. When the stack is almost empty, the busboy may come along with a new stack of plates and push them all into the stack. The stack has only one end and hence requires only one pointer, as shown in Fig. 6.4.4.

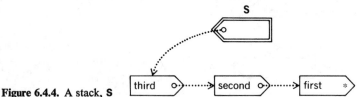

**Figure 6.4.4.** A stack, **S**

***push***     The fact that our list is single ended is reflected in our symbology; we *push* record NEXT and then the record LAST
thus:

$$\text{NEXT} \rightarrow \textbf{S}; \quad \text{LAST} \rightarrow \textbf{S} \tag{6.4.3}$$

***pop***     In Fig. 6.4.5 we see that LAST has been *popped* from **S**
since the pointer from **S** to LAST has been adjusted to point
to NEXT now:

$$\textbf{S} \rightarrow \text{LAST} \tag{6.4.4}$$

The stack is used when the most recently entered record is always to be used first. For instance, think of a list of terms, for example, forks in the

**Figure 6.4.5.** **S** after NEXT and LAST have been *pushed* and showing LAST being *popped*.

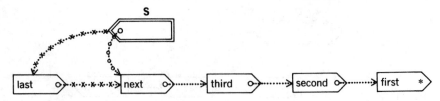

road in the travel instructions a friend gave us. *Push* each one after we use it. When we find we're getting lost, we *pop* the last one and maybe keep *popping* until we're sure of ourselves again. Then we return to *pushing*.

## 6.5 THE SPACE STACK

### The Stack

The space list is a linked list of empty cells headed by a **space head** or space pointer cell. Any cell removed from a list by deletion is appended to the *beginning* of the space list. Whenever a new cell is needed, it is taken from the *beginning* of the space list and the space list is updated to reflect this deletion—the head is pointed to the next available cell.

According to this discipline, cells at the end of the space list rarely are called into use. The most recently added cell is the one that will be withdrawn first. This should be recognized as the principle of last in, first out, or LIFO; the space list is a stack!

The purpose of the space stack is to keep track of available cells and those freed when records are deleted. These latter would otherwise produce *holes* in the list; the cell would be *lost* to us—it could not be reused. Instead, *holes* can now be put onto the space stack. All empty cells or those containing garbage (they need not be cleaned out) are kept track of in the space stack.

In originally setting up the space stack, link together unused sequential cells in the list area. As cells are withdrawn from the space list and later returned thereto, however, the space list begins to wind its way through MEMORY. This is unimportant as long as we have cells available for appending and deleting, and cells arising from records deleted from the file do not get lost in the shuffle; they are *saved* on the space list.

### Interrelation

Several linked lists and a space list may occupy the same linked list area. The usual way to reach a cell in any list is through the head of that list. The head for each linked list bears the name of that list; the head of the space list is called **S** for this discussion.

### Maintaining the Space Stack

The space stack, **S**, contains available cells in **L**. Normally the linked list area, **L**, is referenced for some file **f**, and use is made of the space stack, **S**, in appending new records into a file or when a record is deleted from the file. The details of this are described below. Let us now take a preview of the action required.

***appending***     The action required to append a new record, belle, into file **f** is shown in Fig. 6.5.1. First we reference the space stack and withdraw a new cell, $S_1$. The new record, belle, is placed here. The stack is updated. The record is entered into the list in the proper place.

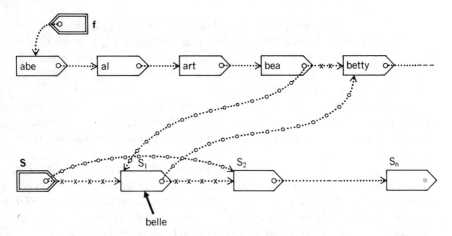

**Figure 6.5.1.** Appending belle to **f**

***deleting***     To delete art from a file the action required is shown in Fig. 6.5.2. The original file, **f**, is restructured so that AL who pointed to ART now points to BEA. Formerly ART would get lost; now ART is pushed into the stack. **S**, which pointed to $S_2$, now points to ART, who, in turn, points to $S_2$.

**Appending**

Now a little more detail about appending. Let us return to Fig. 6.5.1. We notice at the top of the figure the head of file **f**. At the bottom of the figure the head of the space list, **S**, points to available cells that now contain only garbage. They are labeled with reference to the stack: $S_1$, $S_2, \ldots , S_n$. The last cell has a terminal pointer:

$$point\ S_n = * \tag{6.5.1}$$

Before we append belle, we search for her, using *egsearch* and expecting *not* to find her. The search indicates that belle should be attached to BETTY,

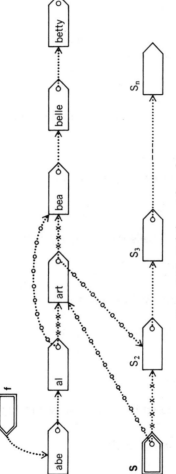

**Figure 6.5.2.** Deleting art from **f**.

175

indicated thus:

$$\text{belle } egsearch \textbf{ f} = \text{BETTY} \tag{6.5.2}$$

Since betty is at BETTY, find the first available cell from the space stack, noting the action thus:

$$point \textbf{ S} = S_1 \tag{6.5.3}$$

Before continuing, update the stack pointer so that the next user can obtain the next cell. This is shown thus:

$$point^2 \textbf{ S} = S_2 \rightarrow \textbf{S} \tag{6.5.4}$$

Now the record is placed into the cell we have just obtained:

$$\text{belle} \rightarrow S_1 \tag{6.5.5}$$

$S_1$ is pointed to BETTY by entering BETTY in its pointer field:

$$\text{BETTY} \rightarrow S_{1P} \tag{6.5.6}$$

Update BEA to point to belle by placing a pointer to the new cell:

$$S_1 \rightarrow \text{BEA}_P \tag{6.5.7}$$

We now have updated the file **f** to show the proper succession:

$$(point \text{ BEA}) = \text{belle}; \quad (point^2 \text{ BEA}) = \text{betty} \tag{6.5.8}$$

### Deletion

To delete art from file **f**, shown in Fig. 6.5.2, search the file and find art:

$$\text{art } egsearch \textbf{ f} = \text{ART} \tag{6.5.9}$$

Note that AL points to ART:

$$point \text{ AL} = \text{ART} \tag{6.5.10}$$

To delete art, have AL point to BEA instead. ART formerly pointed to BEA. The action is then

$$point^2 \text{ AL} = point \text{ ART} = \text{BEA} \rightarrow \text{AL}_P \tag{6.5.11}$$

We are now left with the cell called ART. To get rid of him, he is pushed into the space stack:

$$point \ \mathbf{S} = \mathbf{S_2} \longrightarrow \mathbf{ART_P} \tag{6.5.12}$$

Then point the stack to ART:

$$\mathbf{ART} \longrightarrow \mathbf{S_P} \tag{6.5.13}$$

### 6.6 DOUBLY LINKED LISTS

#### Problem

Using a single linked list isn't a problem when searches are made to find a particular record in the file. The problem arises when we desire to append a new record or delete an old one. In both cases, when we reach a particular cell, we must know its successor and its predecessor; the simple linked list only notes the successor! Let us consider the action for appending only. This requires that we do a search *equal greater* for the key, k:

$$k \ egsearch \ \mathbf{f} \tag{6.6.1}$$

For the simple linked list we require the predecessor of every cell, for we do not know when a match will occur. Suppose we are at W. When we go to its successor, X, we must keep track of W. We have

$$point \ \mathbf{W} = \mathbf{X} \tag{6.6.2}$$

Suppose we set aside two cells to keep track of where we are: NOW and NEXT. We put W into NOW and X into NEXT, and we have

$$(\mathbf{NOW}) = \mathbf{W}; \quad (\mathbf{NEXT}) = \mathbf{X} \tag{6.6.3}$$

During the search we are comparing the key furnished with the key of the record of the cell we are examining. The search continues as long as the relation prevailing is less, thus:

$$(\mathbf{X}) < k \tag{6.6.4}$$

Now we must juggle the pointers:

$$\mathbf{X} \longrightarrow \mathbf{NOW}; point \ \mathbf{X} = \mathbf{Y} \longrightarrow \mathbf{NEXT} \tag{6.6.5}$$

Every time we examine a new record, this juggling is required. The backward pointer eliminates this problem.

### Backward Pointer

Now provide for each cell both a forward pointer, indicated by the subscript P, and a reverse pointer, indicated by the subscript R. The doubly linked list cell has a characteristics shape that is presented in Fig. 6.6.1,

**Figure 6.6.1.** A double linked list cell

also illustrating the location of the pointers and the datum for the cell. The operator *point* extracts the forward pointer thus:

$$point \; E = (E_P) = F \tag{6.6.6}$$

We need a backward operator that extracts the reverse pointer; call it *back*. Then for Fig. 6.6.1 we have

$$back \; E = (E_R) = D \tag{6.6.7}$$

*search*     To search a doubly linked list we proceed exactly as though it were a simple linked list. As we examine a cell and determine that we have not completed our search, we pick up the forward pointer and go on to the next cell.

*appending*     To append a new record to a doubly linked list, we perform first a search *equal greater*. Let us examine this in terms of an example. The example is illustrated in Fig. 6.6.2. In this composite picture we see the list area, **L**, which begins at 3010. In this area we find our file, **f**, and a space list **S**. Both of these are pointed to by their respective heads. It is interesting to see how both the file and the space list wend their way through the space area.

On the right side of the figure we find not only the absolute locations in decimal form but also symbolic names for the locations. To simplify our examination of the appending process, we extract a small portion of the linked list and present it in Fig. 6.6.3. This will illustrate how belle is appended to our present file.

First we do a *search, equal greater*, for belle in the file:

$$belle \; egsearch \; f = BETTY \tag{6.6.8}$$

Then belle should be inserted in the file at the location just before BETTY. Currently BETTY is pointed to BEA since we have

$$back \; BETTY = BEA \tag{6.6.9}$$

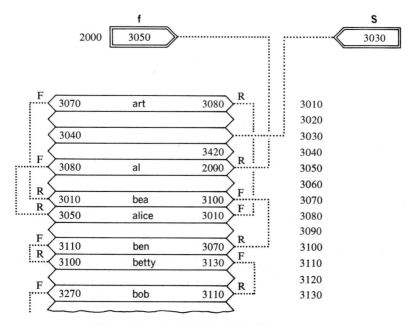

**Figure 6.6.2.** A double linked list file in MEMORY

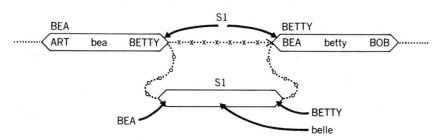

**Figure 6.6.3.** Append belle to f (doubly linked).

Next, go to the space stack and pick up the first cell, indicated thus:

$$\text{point } \mathbf{S} = \text{S1} \tag{6.6.10}$$

This requires an update of the space stack so that **S** now points to S2. This is not illustrated in Fig. 6.6.3 but can be noted in Fig. 6.6.2.

Next the record to be appended is entered into the cell we have just withdrawn:

$$\text{belle} \longrightarrow \text{S1} \tag{6.6.11}$$

Now the pointers in S1 are entered so that the cell points forward to BETTY and backward to BEA:

$$\text{BETTY} \rightarrow \text{S1}_P; \qquad \text{BEA} \rightarrow \text{S1}_R \qquad\qquad (6.6.12)$$

Now both BEA and BETTY must be updated: BEA should point forward to S1 instead of BETTY; BETTY should point backward to S1:

$$\text{S1} \rightarrow \text{BEA}_P; \qquad \text{S1} \rightarrow \text{BETTY}_R \qquad\qquad (6.6.13)$$

Thus four pointers must be entered:

- Two new pointers are required in the new cell.
- A forward pointer in the new cell's predecessor must be altered.
- A backward pointer in the new cell's successor must be altered.

***pointers***    In Fig. 6.6.2 and 6.6.3 we find pointers from one cell to the next to enable us to go forward in our file. There are also pointers from a cell to its predecessor enabling us to traverse our file in the opposite direction. In some cases it may even be desirable to travel backward through the file. Thus in sequentially connecting two cells of our file there should be two arrows: a forward pointer from one cell to its successor and a backward pointer from the successor to the cell. To simplify the figures, an edge, not an arc, is used.

***deleting***    Deleting is illustrated in Fig. 6.6.4 with respect to art. The first action to be performed is a *search equal* to verify that art is in the file and to find the cell that contains him:

$$\text{art } esearch \text{ f} = \text{ART} \qquad\qquad (6.6.14)$$

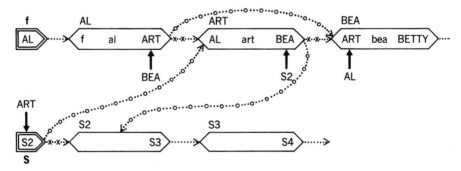

**Figure 6.6.4.** Deleting art from **f** (doubly linked).

Next from the pointers in ART, find his predecessor AL and his successor BEA:

$$back \text{ ART} = \text{AL}; \qquad point \text{ ART} \rightarrow \text{BEA} \qquad\qquad (6.6.15)$$

The forward pointer in AL is made to point to BEA and the reverse pointer in BEA is pointed to AL, so that the list now goes around ART:

$$\textit{point } \text{ART} = \text{BEA} \rightarrow \text{AL}_\text{P}; \qquad \textit{back } \text{ART} = \text{AL} \rightarrow \text{BEA}_\text{R} \quad (6.6.16)$$

This leaves ART hanging. To make him again usable, he is pushed into the space list:

$$\text{ART} \rightarrow \textbf{S} \qquad (6.6.17)$$

ART is connected to the previous top of the list by inserting a pointer to S2 into his forward pointer:

$$\textit{point } \textbf{S} = \text{S2} \rightarrow \text{ART}_\text{P} \qquad (6.6.18)$$

Notice that the space stack is a simple linked list, even though each cell has a place for a backward pointer. This pointer is not needed so long as the cell is in the space stack. Thus ART only has his forward pointer altered. His backward pointer still points to AL and will remain so until AL is put back into use. At that time a new backward pointer will be entered there. Note that the connection between AL and ART has been erased in Fig. 6.6.4: AL no longer points to ART, but ART still points to AL.

### Terminal Pointers

We have used a special symbol, $*$, to designate the last cell (record) in the list (file). It is clear that the designer could have chosen something other than $*$.

Another alternative to a terminal pointer is to create a **circular linked list** by pointing the last cell at the head. It is then possible to enter a linked list at *any* cell and examine the list exhaustively, traversing the list using the pointers. The *head* is like any other pointer. The criterion for completion is *when the pointer in a cell is to the cell at which we entered, it is the last cell.*

We pay a price for circularity—the termination criterion changes from one use to the next—the address of the entry cell. For the **linear linked list** it is always the same, $*$.

### PROBLEMS

6.1 (a) Explain how the linked list (LL) allows an ordered file in an unordered list.
(b) Need LL order be key order in the file? Explain.
(c) Why would we want a LL that is not ordered by key?

6.2 For the LL contained in **L**:
(a) What is the purpose of the *head*?
(b) Why is the head usually kept outside **L**?
(c) Can the head be a cell in **L**? Explain.
(d) How big is the head and why?

6.3  For the operator *point*:
   (a) What does it do when applied to a cell in a LL?
   (b) What does it do when applied to a record in a LL?
   (c) What does it do when applied to the head of a LL?
   (d) What does it do when applied to the last record in the file?
   (e) How do we show that it is applied several times?

6.4  For the operator *esearch*:
   (a) Do the records have to be in the LL in order? Explain.
   (b) What does the operator produce if no record with the desired key is in the LL? Why?
   (c) For each cell *esearch* encounters, several actions are performed. State how to determine
      (i) that the record is found;
      (ii) where to look next;
      (iii) to stop looking.

6.5  In a procedure-oriented language (POL), write
   (a) a procedure POINT to fulfill *point*;
   (b) a procedure ESEARCH that, when supplied with KEY, examines cells of the LL starting at HEAD to fulfill *esearch*; take the exit ABSENT if there is no such record.

6.6  If there is no key order to a LL:
   (a) How do we append record r?
   (b) State the operations required (preferably with operators).
   (c) How do we delete x?

6.7  To append to an ordered LL:
   (a) Why do we employ *egsearch*, not *esearch* or *gsearch* (search for a greater key)?
   (b) What if the record found by *egsearch* has the same key as the new record? Explain.
   (c) Why is it necessary to know the cell in the LL that points to the one found by *egsearch*?

6.8  Write a procedure, EGSEARCH in a POL to search the LL pointed to by HEAD for the place to add a new record with key at KEY. What should be done if there is no record in the LL with key greater than (KEY)? Keep the location of the preceding cell in BEFORE.

6.9  Write a procedure, POP, to find the next cell in the space stack (or pointer), called STACK, and put its address in NEWCELL while updating STACK using POINT.

6.10  Write APPEND using EGSEARCH and POP to add a record now at NEWREC with key at KEY.

6.11  The space pointer or the space stack can be used to keep track of available space in the LL area.

(a) What is the salient advantage of the space stack?

(b) What is the disadvantage of the space pointer?

(c) What is "garbage collection"? With which is it associated? Why is it needed?

6.12 How do we keep variable records in the LL? Explain the needs for

(a) search

(b) append

(c) delete

(d) garbage collection

6.13 (a) Why are fixed format lists so-called?

(b) Explain accessibility for the *queue* and the *stack*.

(c) When would each be employed?

6.14 Write PUSH to update STACK with the cell that is located at OLD using POINT.

6.15 If QUEUE contains two pointers, IPOINT and OPOINT to the input and output of the queue, write the POL procedures IN and OUT to enter a record with address at LOC or retrieve the next record putting its address at LOC.

6.16 Consider the file and space stack, file and space, displayed in Fig. 6.1.

**Figure 6.1.**

Contents

| Address | Key | Data | Pointer |
|---|---|---|---|
| 710 | frank | | 760 |
| 720 | carol | | 840 |
| 730 | joyce | | 820 |
| 740 | howie | | 830 |
| 750 | ben | | 720 |
| 760 | gabe | | 740 |
| 770 | olga | | * |
| 780 | larry | | 850 |
| 790 | nancy | | 770 |
| 800 | ann | | 750 |
| 810 | errol | | 710 |
| 820 | karl | | 780 |
| 830 | irv | | 730 |
| 840 | dave | | 810 |
| 850 | marv | | 790 |
| 860 | | | 870 |
| 870 | | | 880 |
| 880 | | | 890 |
| 890 | | | 900 |
| 900 | | | 910 |
| 910 | | | * |

file [800]

space [860]

We wish to do in this order:
(a) Add gerry;
(b) Add fran;
(c) Delete errol;
(d) Add amy;
(e) Delete gabe;
(f) Add peter.
Show the list after these changes have been made.

6.17 Describe the changes in Fig. 6.1 to make it a circular linked list (CLL).

6.18 What changes are found in the results of Problem 6.16 when Fig. 6.1 is a CLL?

6.19 Redraw Fig. 6.1 as a doubly linked list (DLL) where each entry has address/key/datum/back pointer/forward pointer. How does the head change? Show this.

6.20 Do Problem 6.16 for the list created in Problem 6.19.

6.21 (a) Consider a circular doubly linked list (CDLL). Explain.
(b) State what changes occur in the answer to Problem 6.19 for the CDLL.

# CHAPTER 7

# Branching and Shared Linked Lists

## 7.1 BRANCHING LISTS

### Need

Although simple files for posting are generally linear or serial, other applications require the representation of trees or transitive graphs. An example of this is a parts breakdown list for a large mechanism such as an automobile. Consider the car first to consist of major assemblies; these in turn might be divided into smaller subassemblies, and so on, until finally we list single parts. This structure is perfect for representation by a tree or sometimes even a transitive graph.

Another example is an index system for books with references and cross-references. This is again a non-linear graph.

Graphs of trees contain nodes that are rootlike. One arc enters the node but several arcs leave. The linked list can represent the tree but we need a way to represent this type of vertex. The branch cell, which we now examine, does this. Leaflike vertices are also needed when shared sublists occur. These are needed for transitive graphs, a topic postponed until Section 7.6.

### Branch Cell

The normal linked list cell has only one forward pointer in it and therefore can serve only as a simple (connecting) vertex. A **branch cell** can point to at least *two* other cells. There are several ways that we could set up the branch cell:

- It may or may not contain a datum.
- It may contain exactly two or more than two pointers.
- It may be the same size as a record cell or it may be smaller.

Let us examine a few of these combinations.

### Uniform Record Format

With this technique, all linked list cells contain an equal number of data fields, enough to make up a full record; all cells are the same size; pointer fields are provided to at least two other cells. Only one of these fields is used in a normal record at a simple node; whereas two pointer fields are used when the cell is at a rootlike node.

There are obvious advantages to having all cells the same size:

- They can all come from the same space list.
- *All* cells may contain a datum.
- The search algorithm may be consistent.

There are disadvantages, among which are the following:

- There is the extra pointer field in nonbranch cells.
- Although the search algorithm is uniform, it takes longer.
- For nodes with incidence greater than three we have to use a record solely as a pointer, eliminating its data-carrying possibilities entirely and wasting MEMORY space.

### Bifurcating Branches

When a cell contains exactly two pointers, it is a **bifurcating branch cell**. Representation of trees with multibranch nodes require more branch cells than if several pointers were carried in each branch cell but it does make for a uniform search policy.

### Data Branch Cells

A linked list using **bifurcating data branch cells** is presented pictorially in Fig. 7.1.1. Each multibyte record is compressed in the figure for simplicity in representing all the data fields by a single symbol. Then at

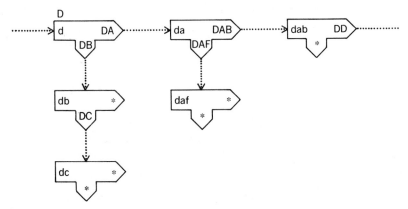

**Fig. 7.1.1.** A linked list using bifurcating data branch cells

D we have the data field, d, a primary pointer, DA, and a secondary pointer, DB. This is a data cell and there is also a secondary pointer. In cells with only a single pointer such as DB and DAB, there is still a secondary pointer field, and it must be filled with something; * indicates that this field is void.

In Fig. 7.1.2 we see the tree from which the list of Fig. 7.1.1 was derived.

The difficulty with this technique is shown in Fig. 7.1.3 and 7.1.4. The former figure can easily be duplicated using the data branch cells described above: a cell at J contains a datum j and two pointers, R and K.

When it is necessary to represent the graph of Fig. 7.1.4, we find it impossi-

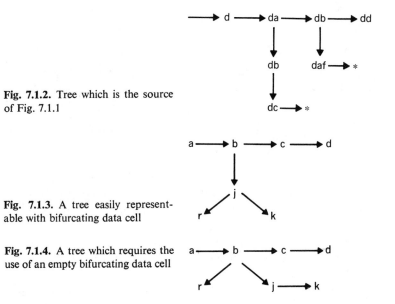

**Fig. 7.1.2.** Tree which is the source of Fig. 7.1.1

**Fig. 7.1.3.** A tree easily representable with bifurcating data cell

**Fig. 7.1.4.** A tree which requires the use of an empty bifurcating data cell

ble to represent the three-way split from B, respectively, to C, J, and R, without using an *empty* data cell. The reader may try, noting that this is the only expedient that will work. In Fig. 7.1.5, the solid circle represents a cell that contains no data, only two branch pointers.

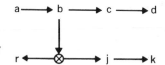

**Fig. 7.1.5** The circle is for the empty data cell.

### Branch-Only Cells

We can set up **branch-only cells** without data. These are tagged as branch cells. They might have two or more pointers, as shown in Fig. 7.1.5. These pointers may point to data cells or to other branch-only cells. Since their function is so limited, it is obvious that these cells can be smaller than data records. Then we need to tag all cells, however, to distinguish between data and branch cells. A data cell contains

- a tag;
- a datum or a data pointer;
- a link pointer.

Hereafter in examples which follow we shall use **bifurcating branch-only cells** which contain

- a tag;
- two link pointers.

Since branch cells may be of different size than record cells, it is advisable, if not mandatory, to have a separate **branch space list** (say, **B**) for available branch cells.

### Secondary Pointer Operator

Consider a branch pointer at B. It still has a primary or forward pointer to NEXT, say. As before, it is extracted by the operator *point*. But there is also a secondary pointer to BRANCH, say. How do we get to it?

We postulate a **secondary pointer operator** *branch*, which when applied to a branch cell produces the secondary pointer contained therein. Thus:

$$point \ B = \text{NEXT}; \qquad branch \ B = \text{BRANCH} \qquad (7.1.1)$$

Of course, when applied to a data-only cell, *branch* will come up with nothing (∗). If D is a data cell that points only to MORE, then

$$point \ \mathsf{D} = \mathsf{MORE}; \quad branch \ \mathsf{D} = * \qquad (7.1.2)$$

## 7.2 CONSTRUCTING BRANCH LISTS

### Trees

Linked lists with the aid of branch cells are now able to hold any directed tree. It is the intent of this section to indicate how, given a particular tree, we can construct a branch list that represents it. We shall use the tree of Fig. 7.2.1 as an example for conversion.

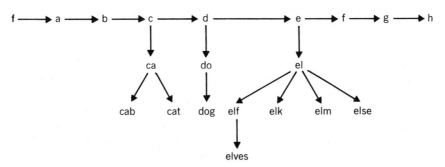

**Fig. 7.2.1.** A tree with several branches

***bifurcating***    A simple extract of Fig. 7.2.1 appears in Fig. 7.2.2, where
***branch cell***    e points to f and also to el. We use nondata binary branch
cells; e is in a data cell that cannot branch so it points to a branch cell. This is again presented in Fig. 7.2.3, where the large dot represents a branch cell.

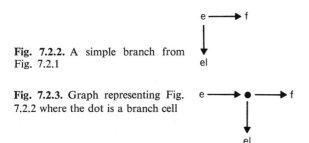

**Fig. 7.2.2.** A simple branch from Fig. 7.2.1

**Fig. 7.2.3.** Graph representing Fig. 7.2.2 where the dot is a branch cell

***implemen-***     In Fig. 7.2.4 the cell C points to B1, which is a branch cell
***tation***        and as such it contains two pointers:

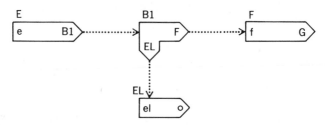

**Fig. 7.2.4.** Linked list representation of Fig. 7.2.2

- The primary pointer, $B1_P$ points to the next element in the list, f, contained in the cell, F.
- The secondary pointer, $B1_S$, contains EL to point to the cell called EL containing the datum, el.

***multiway***     When we come to the portion of Fig. 7.2.1 that has been
***branch***       extracted and appears as Fig. 7.2.5, we find that el is a root-
                   like node from which four branches emanate. It would be
convenient if one branch cell could provide pointers for all these branches.
That case is presented in Fig. 7.2.6. Here el points to the large dot that represents a four-way branch cell and takes us in the four directions shown. A
diagrammatic implementation of this appears as Fig. 7.2.7.

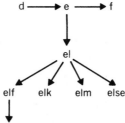

**Fig. 7.2.5.** A multibranch node

**Fig. 7.2.6.** A multibranch cell easily
represents a multibranch node in
graph form.

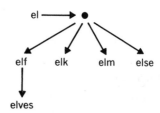

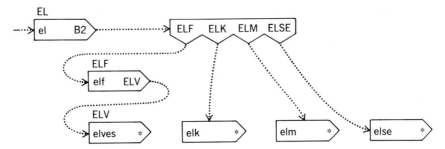

**Fig. 7.2.7.**  A multibranch cell in a linked list

***using***
***multiple***
***binary***
***branch cells***

If we restrict ourselves to binary branch cells, then the tree extract of Fig. 7.2.5 can now be replaced by that of Fig. 7.2.8. We need three binary branch cells,

$$B2, B3, \text{ and } B4,$$

to provide the realization, which is then presented in Fig. 7.2.9.

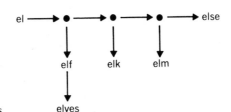

**Fig. 7.2.8.**  Using bifurcating nodes

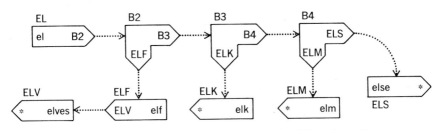

**Fig. 7.2.9.**  Linked list with a multibranch node using bifurcating cells

**User's Task**

Suppose that the user has the directed tree presented in Fig. 7.2.1 that is to be placed in linked list format using binary branch cells. What does he do?

***restructured***
***tree***    In Fig. 7.2.10 we see how the user might restructure this tree, inserting bifurcating cells where they are necessary.

A program to operate on this reconstituted tree will still be difficult to construct, so we ask the user to take things one step further.

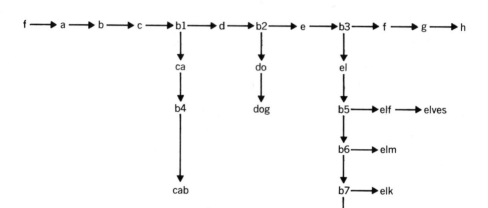

**Fig. 7.2.10.** A full graph showing branch cells to Fig. 7.2.1

***pruning the***
***branches***    The user determines the **trunk** (or main line) of his tree, which begins with the root. The trunk includes the branch cells but without any branches, the trunk of the graph in Fig. 7.2.10 appears at the top of Fig. 7.2.11.

For each branching node in the trunk the user now enters the corresponding branch. The branch in turn may include a branch cell, but the secondary

**Fig. 7.2.11.** The trunk and separate branches of Fig. 7.2.10

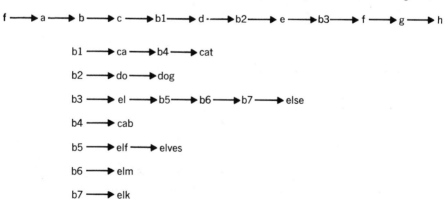

branch from that cell too has been trimmed and only the primary branch indicated.

Branch nodes that appear in branches are part of a branch graph. Thus in Fig. 7.2.11 all branch nodes are accounted for whether they appear in the trunk or in secondary, tertiary, etc., branches.

### Trunk Entry

First place the trunk of the branching linked list into the list area. Recall that the trunk consists of those cells that use only the primary pointers and, for the example, is shown at the top of Fig. 7.2.11.

To construct the trunk, simply enter records as they appear in the input stream, obtaining cells for the records using the space pointer or space stack and causing each to point to the next. When a branch cell is required, it is obtained from the branch space stack if one is provided. For instance, after c in the trunk tree we find b1 and d. An empty branch and data cell, respectively, are required for them; this portion of the list appears in Fig. 7.2.12. Notice that the secondary pointer in B1 is left blank; we don't have any information about branches yet.

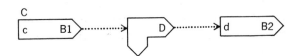

**Fig. 7.2.12.** How b1 is replaced by B1.

*branch queue*    After the trunk is constructed, add branches one at a time. To locate each branch, search the file F serially, until a branch cell is encountered.

An alternative and perhaps superior method is illustrated in Fig. 7.2.13. During construction as each branch cell for the trunk is selected, we *enqueue* an auxiliary cell on the queue, illustrated in Fig. 7.2.13. Each queue cell (for example, Q1) contains

- the name of the branch cell (b1);

**Fig. 7.2.13.** A queue to keep track of incompletely filled branch cells

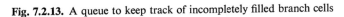
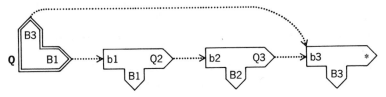

- a pointer to the next queue cell (Q2);
- the location of the branch cell (B1).

The queue for the trunk of the example appears in its entirety in Fig. 7.2.13. The name of the queue is **Q**; since it is a queue, its head has two pointers, one to the front and one to the back of the queue.

### Adding Branches

There is further input required that describes each branch of our tree, as illustrated in Fig. 7.2.11. This information is processed from the input stream. First the branch labeled b1 occurs; if the branch queue method is used, *dequeue* **Q** to obtain an entry for the branch labeled b1 together with its location, noted as B1.

Next, cells are acquired for the remainder of the branch and the pointers are filled in, as illustrated in Fig. 7.2.14:

- The cell, CA, is obtained and filled, and a secondary pointer to it is inserted in B1.
- A branch cell B4, is obtained and a pointer to it is inserted in CA.
- A cell, CAT, is obtained and a primary pointer to it is inserted in B4.
- Since there are no more items on this branch, a terminal pointer ∗ is entered in CAT.

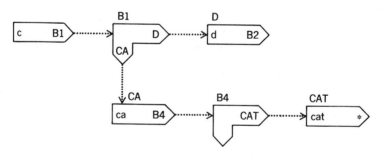

**Fig. 7.2.14.** Using the branch graph to set up the first branch

As the branch, b1, was entered into the list we note the following:

- The cell B1 was completely filled in and the corresponding cell Q1 was removed from **Q**;
- As a new branch cell B4 was entered into the new branch, a new entry Q4 was placed on **Q** to locate it and fill it in later.

Figure 7.2.15 shows how the queue, **Q**, appears after the branch, b1, has been entered. Notice that the cell, Q1 is missing and a new cell, Q4, has been added.

Processing of all the branches continues thus since all branch cells should contain two pointers. When the list is complete, we should find that **Q** is empty.

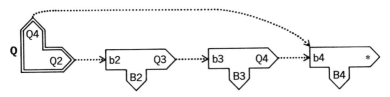

**Fig. 7.2.15.** The queue **Q** after the first branch has been filled in

## 7.3 SEARCHING BRANCH LISTS

There are two ways to search branch lists:

- An algorithm may be employed that indicates when to stop the search.
- The list may be searched exhaustively, in its entirety.

### Algorithm

To use an algorithm to recover a particular record, it is helpful to know the algorithm used for entering the records into the list originally; it may be possible to adopt it without change. This is the means for entering cells into the main list and also the rule for the order in which cells appear in the branches. Further rules may exist for entry that are not used during the search process.

*example*     Here is an example of rules adhered to in constructing a tree for later record retrieval:

1. Records appear in ascending order in the trunk.
2. Records on a branch from the trunk have keys
   - higher than the record on the trunk from which the branch comes;
   - lower than the successor record on the trunk.
3. The same rule applies between **twigs** (branches from branches) and branches that they grow on.

**tree**     Figure 7.3.1 shows a tree that fulfills the rules above. All
the branches from the node c have keys that are less than
the next node d; that is, they all begin with the letter c and consist of two or
more letters, so they are greater than c but smaller than d.

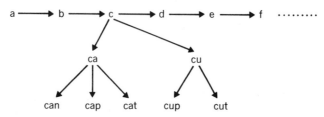

**Fig. 7.3.1.** A portion of a tree for algorithm search

**binary**     Figure 7.3.2 shows a portion of the tree in Fig. 7.3.1 con-
**branching**     verted into a binary branching list where large dots indicate
**list**     branch cells. Figure 7.3.3 shows the arrangement of the cells
in the bifurcating list to reproduce the tree of Fig. 7.3.2.

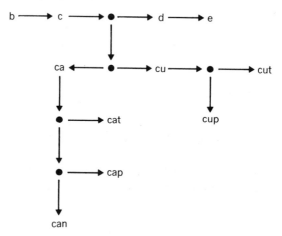

**Fig. 7.3.2.** A bifurcating branch tree for Fig. 7.3.1

**search**     Let us find the record, cat, where the search is illustrated
by dashed lines in Fig. 7.3.3. The numbers in parentheses
in the text apply to the numbers on the search route.
First, perform a *search equal qreater* along the trunk:

$$\text{cat } egsearch = D \tag{7.3.1}$$

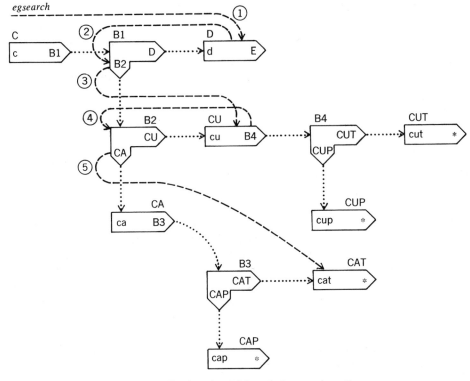

**Fig. 7.3.3.** A bifurcating branch list for Fig. 7.3.1 and the search path for cat using the algorithm

Cell **D** contains the record d (1); we have gone too far and now backtrack. If the previous cell is a data cell, then the desired record is missing. In the figure the previous cell is the branch cell, **B1**. Extract its secondary pointer and proceed from here (2):

$$back \cdot D = B1; \qquad branch\ B1 = B2 \qquad (7.3.2)$$

or

$$branch\ back\ D = B2 \qquad (7.3.3)$$

We now perform a search along that branch (3):

$$cat\ egsearch\ B2 = CU \qquad (7.3.4)$$

From cell **CU** we find again that we have gone too far and backtrack to the previous branch (4). Extract the secondary pointer and proceed from there:

$$branch\ back\ CU = CA \qquad (7.3.5)$$

Traverse this branch (5) to CAT to find the record of interest:

$$\text{cat } egsearch \text{ CA} = \text{CAT}; \quad (\text{cat } esearch \text{ CA}) = \text{cat} \quad (7.3.6)$$

The reader should carry on a similar search for the record car, evidently missing from the list.

### Exhaustively

When there is no rigorous method for finding records in the branching linked list, we search through the entire list exhaustively to be sure that we get to every record in the file somehow. This should be fairly easy to do since *all* branch points are bifurcating. Hence if we leave a node in one direction, we should eventually return to that node to leave in the opposite direction. We could try a scheme where we would go along the *trunk* first when we examine the file. Later we would go up to the first node, take the secondary path, and continue, using primary pointers and so forth. The trouble is, many cells would be examined several times.

There are many techniques for reducing the number of cells examined, and only one is presented here. The trick is to use an auxiliary queue or stack to keep track of branch points. Here we use a queue: since it is an auxiliary Queue for Searching, call it **QS**.

*the tree*      Presented in Fig. 7.3.4 is the same tree that appeared as Fig. 7.2.10, with the addition of search paths for exhaustive search. Numbers on the tree are cued in with the description.

*technique*      First traverse the trunk using the primary pointers. *Enqueue* each node encountered into **QS**. We could enter the location of the branch cell. It is simpler to enter the secondary pointer extracted from the branch cell. For instance, in Fig. 7.3.4 at B1 the first entry into **QS** is the secondary pointer from B1, indicated symbolically thus:

$$branch \text{ B1} = \text{CA} \rightarrow \textbf{QS} \quad (7.3.7)$$

After examination of the trunk, the queue contains a number of entries, as at the top of Fig. 7.3.5. Obtain the first of these and continue to traverse cells pointed to by primary pointers along this branch. As we encounter more branch cells, secondary pointers are extracted and entered in the queue. This continues for all the nodes along the branch. After each branch is examined, *dequeue* **QS** until it is empty—then the entire file has been searched.

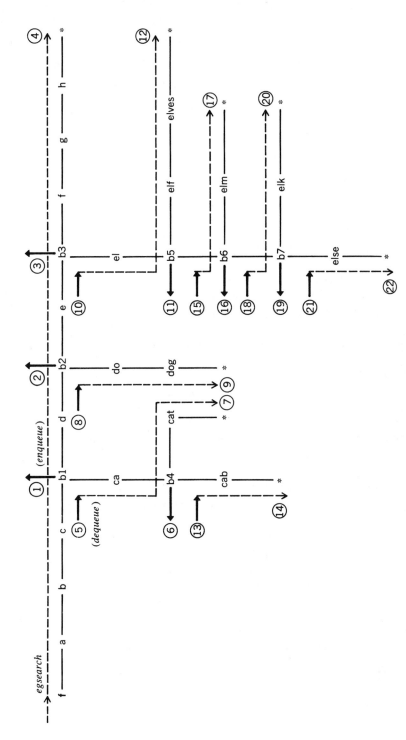

**Fig. 7.3.4.** Search path for the tree of Fig. 7.2.10

199

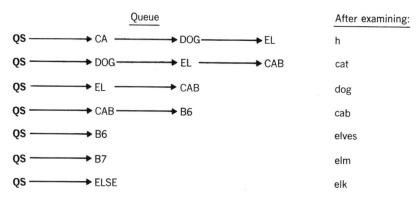

Fig. 7.3.5. QS for the search in Fig. 7.3.4

**Example**

Return to Fig. 7.3.4 and traverse the trunk. We encounter three branch cells and *enqueue* the secondary pointers from each of these in order to **QS** (1, 2, 3). The contents of **QS** is shown in Fig. 7.3.5. At the end of the trunk we find that H has a terminal pointer (4). *Dequeue* **QS** to obtain CA (5), which points to B4, and enter the secondary pointer into **QS** (6). Continue until we hit the terminal pointer in CAT (7). **QS** now appears as in the second line of Fig. 7.3.5. *Dequeue* DOG (8), which has a terminal pointer (9).

Continue by examining the complicated branch at B3 (10). The rest of the search is easy to follow.

The search terminates at ELSE (21), which contains a terminal pointer (22). When **QS** is referenced, it is found to be empty, thus guaranteeing that there are no more branch points for which secondary branches remain to be examined.

## 7.4 ALTERING BRANCH LISTS

For branch lists we may also wish to append or delete elements. When this action is performed, however, new branches may be created or old branches destroyed. This leads to the creation or annihilation of branching links. Hence whenever alteration occurs, we should check the branch link that may precede the altered elements.

### Deletion

Deletion of a normal cell sometimes takes a branch link with it. It is hard to anticipate when this may happen. For instance, in Fig.

7.4.1 the deletion of ca (1), although it is at a branching node, doesn't take a branch with it; whereas the deletion of cat (2), which is a leaf, should take the branch link with it.

This is better seen in Fig. 7.4.2. Here b1 is required between c and b to make the branch to ca possible. When ca is removed, the branch still remains and both b1 and b4 are necessary.

The actions required to alter the list are shown in Fig. 7.4.3. To remove

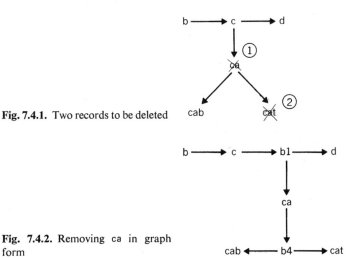

**Fig. 7.4.1.** Two records to be deleted

**Fig. 7.4.2.** Removing ca in graph form

**Fig. 7.4.3.** Removing ca from the linked list

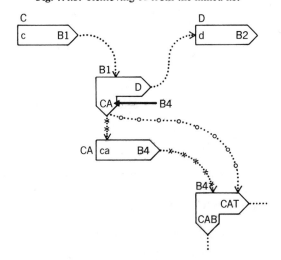

ca, B4 is entered into B1 as its secondary pointer. The secondary pointer now points to the next branch point, having been detoured around CA. If we use a space stack, CA is returned to it.

Now suppose that we wish to delete cat, leaving ca. The action to be performed is shown in graph form in Fig. 7.4.4. The branch node b4 is necessary so that ca could point to both cab and cat. Since we are deleting cat, b4 is not necessary either.

The action performed, shown in Fig. 7.4.5, releases CAT and B4. The former can be returned to the space stack. If a branch space stack is also kept, B4 is returned to it. The result is shown in Fig. 7.4.6.

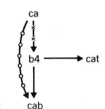

**Fig. 7.4.4.** Deleting cat graphically

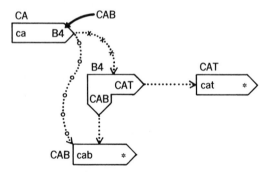

**Fig. 7.4.5.** Removing CAT also gets rid of B4

**Fig. 7.4.6.** After CAT and B4 are gone

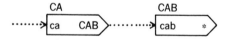

### Appending

To insert a new record in a file, all that is needed is a new cell and proper pointer adjustments. It is only when a new branch is created that additional branch cells are required.

*example*    In Fig. 7.4.7 two new records are to be added, fan and fat.
They create two new branches and two new branch cells are required.

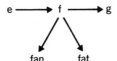

**Fig. 7.4.7.** Adding fan and fat

In Fig. 7.4.8 we see a graph that indicates with dots the two new branch cells. Such a graph conveys where new branches, branch cells, and records belong in the file. With enough information, a program can update the file. The result of the action shown graphically, appears as Fig. 7.4.9. A list diagram illustrating this is Fig. 7.4.10.

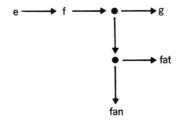

**Fig. 7.4.8.** Two branch nodes are required to add fan and fat.

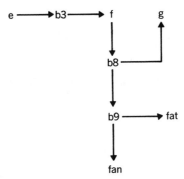

**Fig. 7.4.9.** We use b8 and b9 to add fan and fat.

## 7.5 DOUBLY LINKED BRANCH LISTS

### Need

Doubly linked branch lists improve and simplify the search algorithm. We have seen how *egsearch* requires for a *greater* comparison that we retrace our steps to the predecessor branch node. Retracing is facilitated when cells contain a reverse pointer.

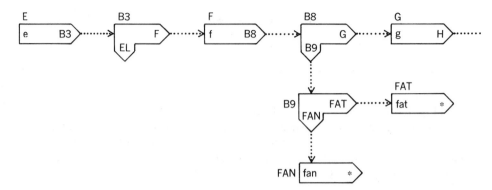

**Fig. 7.4.10.** Altogether we have to add B8, B9, FAN and FAT.

Another need arises when branch lists are subject to revision and we append new records. *Egsearch* finds the proper place to append the new record between the designated cell and its predecessor. Finding the predecessor is easy with the reverse pointer.

### The List

The list, as before, consists of two kinds of cells. The data cell, illustrated in Fig. 7.5.1, contains both a forward and reverse pointer. The branch cell now has three pointers. Figure 7.5.1 portrays cell B3, which contains the forward pointer, $B3_F$, namely F. The secondary pointer, $B3_S$, points to EL. The reverse pointer, $B3_R$, points back to E.

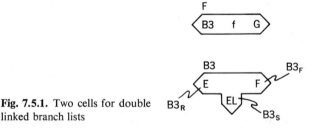

**Fig. 7.5.1.** Two cells for double linked branch lists

To illustrate how these cells are used, a portion of Fig. 7.2.1 is now presented as Fig. 7.5.2. The branching nodes to be illustrated are b3, b5, and b6.

Figure 7.5.3 shows the list form of the graph in Fig. 7.5.2. It should be clear how the branch cells are used.

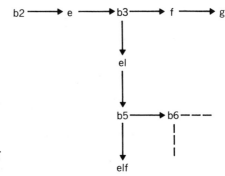

**Fig. 7.5.2.** A portion of the tree of
Fig. 7.2.1 showing nodes

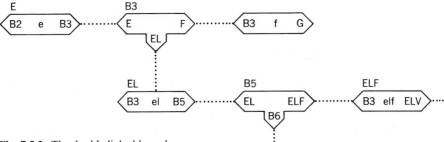

**Fig. 7.5.3.** The double linked branch

### Deleting

The method for deleting a data cell was examined in Section 6.6, and the action is identical for doubly linked branch lists. The procedure is just as simple when the data cell deleted also takes with it a branch cell. Figure 7.5.4 provides a graph of how cat is deleted. The list form of this appears as Fig. 7.5.5. To delete the record cat, the cell CAT is withdrawn from the list. As this is done, the reverse pointer in CAT is followed to B4, which is tagged as a branch cell. The forward pointer of B4 would now become a terminal pointer. A branch cell with only a secondary pointer (or only a primary pointer) really does no work and can be removed—B4 is also removed.

**Fig. 7.5.4.** Deleting cat

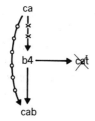

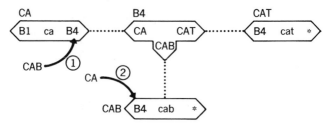

**Fig. 7.5.5.** Cat takes CAT and B4 with it.

To update the list, the secondary pointer from B4, CAB, is placed in the forward pointer of CA (1); now CA points to CAB. The reverse pointer in CAB still points back to B4, however. It is altered to point to CA (2).

### Appending

Appending a new data cell that does not create new branches is a simple matter. We need only be sure, as explained in Section 6.6, that both forward and reverse pointers are properly reset. The novelty occurs when new branches are created. An example is presented in graph form as Fig. 7.5.6. Here two new records, fan and fat, are added, both of which create

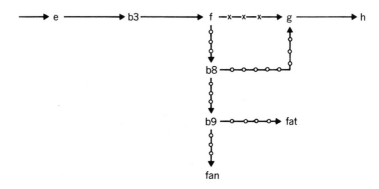

**Fig. 7.5.6.** Adding fat and fan

new branches and hence require two new branch cells. The portion of the list resulting from these additions is presented in Fig. 7.5.7. One thing to note is that F, which formerly pointed to G, now has the branch cell, B8, intervening. The forward pointer in F and the reverse pointer in G are now both set to B8. As we would expect, the forward pointer of B8 is to G, and the reverse pointer is to F; the secondary pointer is to B9 to provide the two new branches.

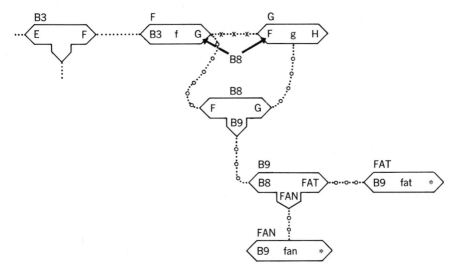

**Fig. 7.5.7.** Fat and fan bring FAT, FAN, B8, and B9.

## 7.6 SHARED SUBLISTS

### Introduction

A subfile corresponds to a branch in a graph. Why not let several users reach a single subfile? Or, to put it graphically, why not let the same branch be shared by other branches in the graph?

Figure 7.6.1 shows visually what might be required here. We see a sublist sh that we would like to be accessible to two nodes of our graph, x and y.

**Fig. 7.6.1.** A shared sublist, sh

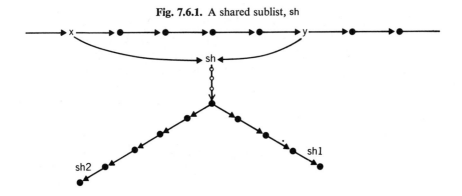

sh does not *belong to* x and y but it is accessible to both. Notice also that sh consists of two branches in its own right, sh1 and sh2.

Once having left the trunk at x or y we pass into the branch and terminate, using either sh1 or sh2. Thus it is not possible to return to the trunk after having taken the branch route.

Notice that the graph in Fig. 7.6.1 contains a cycle, xyshx, but not a loop, for it isn't possible to travel from x back to x.

### Linked List Requirements

To put the graph of Fig. 7.6.1 into linked list form, nothing new is required. We see an implementation in Fig. 7.6.2.

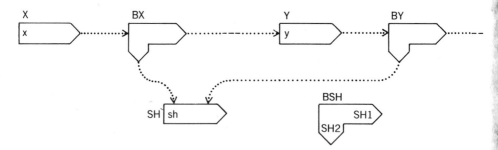

**Fig. 7.6.2.** Linked list form of a shared sublist

Both x and y are branching nodes so a branch node is required for each—find BX and BY in the figure. What is new is that the secondary pointers of both BX and BY point to SH since they share this branch.

### Practical Use

We have developed the linked list as a tool for expressing not only linear lists but trees and transitive graphs, structures not otherwise representable. Shared subfiles are particularly valuable in conveying transitive graphs such as parts lists. Consider a large mechanism that consists of several assemblies. Along the trunk of the graph each node represents a different assembly. If it is nonbranching, it is a single part; a branch *is* the assembly. Other nodes appear along the branch for parts and subassemblies that make up the assembly.

Effective designers economize and use the same subassembly for several assemblies if this is possible. Thus in the total graph for the mechanism we may find assemblies that contain common subassemblies. The subassembly is represented as a branch that is *shared* by the assemblies.

### Example

Figure 7.6.3 shows a somewhat more complicated example where the subfile, sub1, is shared by several nodes in the trunk, namely assy1, assy2, and assy4. Notice in the figure that sub2 is also a shared sublist.

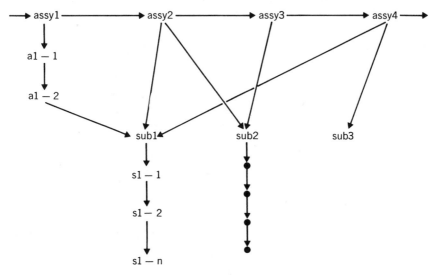

**Fig. 7.6.3.** Graph of example with two shared sublists

Figure 7.6.4 shows how the graph of Fig. 7.6.3 is implemented. As we would expect, each branch node of the trunk requires its own branch cell. We can follow how the branch from assy4 takes us through component parts to the sublist at SUB1.

*double links*     We have seen the usefulness of doubly linked lists where backward pointers permit us to retrace our steps. For a shared sublist, several nodes point to one sublist. Therefore, instead of one reverse pointer, we need several. In Fig. 7.6.4, SUB1 would have to point backward to B4-1, B2-1, and B6-1.

This is not so farfetched. In fact when it is necessary to "implode" instead of "explode" a parts list, we want to go back from a subassembly to find all assemblies that use it. In this case we provide a large reverse pointer field that can hold as many reverse pointers as we might need.

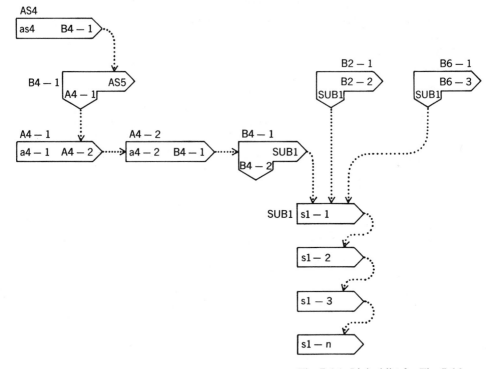

**Fig. 7.6.4.** Linked list for Fig. 7.6.3

### Traversing Shared Sublists

Obviously the pointers in the linked list provide the means for going from one data cell to another. If there is an algorithm for taking us to the desired cell or group of cells, then there should be no problem. It is when we want to examine the list exhaustively that the problem arises. Since there are several ways in which we can reach a single shared sublist, we run the risk of examining the same sublist several times. This will not hurt but it does consume time. Auxiliary stacks or queues may keep track of sublists we have already examined, thus saving us this trouble. Still it is best to use the algorithmic approach.

### PROBLEMS

7.1  (a) What is a *branch cell*?
    (b) Why does it contain two or more pointers?

(c)  For what kind of graphs are they needed?

7.2  What are the objections to the following?
(a)  Every cell is capable of branching.
(b)  Every branch cell has more than two pointers.
(c)  Cells may be of variable size and have a variable number of pointers.

7.3  What is meant by a *nondata bifurcating branch cell* (NBB cell)? Give its advantages and disadvantages.

7.4  Why is it impossible to represent a tree as a sequential file without supplementary pointers?

7.5  For the NBB cell in Problem 7.3:
(a)  Why may it be smaller than a data cell?
(b)  Why must it be tagged?
(c)  Why is there a special space stack **B** for them?

7.6  For the operator, *branch*:
(a)  What is its purpose?
(b)  What does it produce from a data cell?
(c)  Could *branch* produce that result from an NBB cell? Explain.

7.7  A node in a tree points to five other nodes; how many NBB cells are needed to reproduce it? Generalize this.

7.8  What is the *trunk* of a tree? Is it unique? Operationally, how is the trunk defined for the LL that represents it?

7.9  The text and (7.3.1) through (7.3.6) explain how to search an *ordered transitive LL* (OTLL). Put this into the form of an algorithm.

7.10  The LL being searched in Problem 7.9 is called on OTLL. What does that mean? Why is this a requirement for the algorithm to work?

7.11  (a)  For the exhaustive search (7.3.3), is the *egsearch* or *esearch* operator used?
(b)  What requirement for order is placed on the LL?
(c)  Why is **QS** required here but not for the OTLL?
(d)  When a branch cell is reached:
(i)  How does search continue?
(ii)  How is **QS** used?

7.12  To delete a record from a branching LL (BLL):
(a)  What happens to the cell it occupies?
(b)  Why *may* a branch cell go with it?
(c)  What is the criterion for deleting a branch cell too?

7.13  To append to a BLL:
(a)  When is only a data cell added?
(b)  When is a branch cell added too?
(c)  What is the criterion for this?

7.14  (a)  What is a doubly linked branching linked list (DBLL)?

    (b) What pointers are required in a data cell?

    (c) What pointers are required in a NBB cell?

7.15 Develop an algorithm for

    (a) searching the ordered DBLL;

    (b) searching exhaustively the unordered DBLL with a queue;

    (c) appending a new record in an existing branch with a space list **S**;

    (d) appending a new record in a new branch with a branch list **B** and a space list **S**;

    (e) deleting for part c;

    (f) deleting for part d.

7.16 (a) The shared branching linked list (SBLL) extends the ability to represent graphs from trees to what other structures?

    (b) What graph structures are *not* representable as LL's?

    (c) May the SBLL contain cycles? loops? Explain.

7.17 Explain at least two applications of the SBLL.

7.18 For the graph that the SBLL represents there are several kinds of vertices. Explain how they are conveyed in the SBLL:

    (a) root,                  (d) leaflike node,

    (b) leaf,                  (e) simple connecting node,

    (c) rootlike node,       (f) multiple connecting node.

7.19 As for Problem 7.18a through f, examine the skewed branching *doubly* linked list (SBDLL).

7.20 Using one or more auxiliary queues or stacks, provide an algorithm to examine exhaustively a SBLL.

# CHAPTER 8

# *Directories*

## 8.1 NEIGHBORHOOD LOCATABILITY

We would like to find a list structure that would permit easy search and at the same time allow for changes—appending and deleting. Neighborhood locatability has both advantages but incurs some disadvantages.

I have coined the term **neighborhood locatability** because it is expressive of a group of methods having a similar characteristic: the technique leads us immediately to a **neighborhood** in the list and localizes our search to this neighborhood. The advantage is apparent—we immediately cut down the range of our search to a neighborhood. The disadvantages are that further search may be required and that appending and deleting require more work than for the linked list.

### Searches

We have discussed several search methods that are now reviewed.

For a **serial search** we examine each list cell in the order in which it is situated in MEMORY. We search a linked list in the order in which its cells are linked together. Our motto is "keep looking." If the file is ordered, we can terminate search before all the cells in the list have been examined. **Binary search** (of an ordered list that is preferably dense) homes in to a smaller and smaller area within which the desired record lies. This method requires a much smaller number of looks than serial search, especially for large lists; however, the calculations made to home in on the target consume time.

Search in the neighborhood locatable list immediately brings us to the correct area. The method employed depends on the nature of the list.

### Problems

Associated with neighborhood locatability are several problems, itemized below and discussed in the rest of this section:

- **transformation**: how do we convert from the desired key value to the neighborhood in MEMORY where that record will be found?
- **search**: how do we find the desired record in the neighborhood or determine that the record is absent from our list?
- **changes**: how do we add new records to the list or delete unwanted records from the list?
- **overflow**: if the area assigned to lists or sublists is all used up, how do we obtain an assigned new area?

### Transformations

Given the key, how do we find the neighborhood?

*directory*     By the **directory** or **table of contents** method, we first divide our list into sublists. We then prepare a table of contents wherein each sublist has an entry. We use a key to enter this table of contents. We find the proper entry that takes us to the sublist—the neighborhood. Other information in the entry tells us how to search the sublist. The directory method is described in Sections 8.2, 8.3, and 8.4.

*multiple*      This method applies a directory to the directory. That is,
*directory*     with a very large list, we find the directory is also large. We make a directory for the directory. The master directory takes us to a subdirectory, a portion of the original directory. This, in turn, takes us to the sublist where the record may be found. This is described in Sections 8.5 and 8.6.

*processing*    Instead of using a directory, we may operate directly on the
*the key*       key to find a location in MEMORY that is the beginning of our search.

We map the key into a cell; this cell is the start of a neighborhood. From this cell we generally move in only one direction, forward—although bidirectional searching allows us to move in either direction according to what happens when we look at the first record. The operation that we perform on the key is the mapping. Chapter 9 is devoted to mapping.

## 8.2 THE TOTAL DIRECTORY

### Why Total?

A **total directory system** is so-called because the directory is provided with one entry for each record. This makes finding a record easier, especially when

- records are of variable length;
- the list area is on an auxiliary medium for which a linear search is time-consuming.

### The Directory

A total directory presented in Fig. 8.2.1 consists of one entry for each record in the list. **D** is the name of the directory, and directory

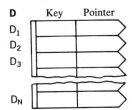

**Fig. 8.2.1.** The total directory entry, one for each record, contains the record key and a pointer to the record.

cells are subscripted in order of their appearance in the directory, so that $D_1$ is the first and $D_N$ is the last. The entry consists of two parts:

- The key in the directory entry is the key of a record in the file.
- The pointer is the location of that record in the list.

The directory is 1 : 1—that is, there is one entry for each record in the file and for each record there is just one entry in the directory.

It should be noted that the record may reside in the COMPUTER MEMORY or in some auxiliary medium. For the former the entry provides the byte location of the record; for the latter the location may be the relative or absolute position of the record in the file or on the mechanism, respectively. Often the relative block and track number of the record is found in the entry pointer.

The pointer operator, *point,* that we have employed earlier for linked lists will also extract a pointer from a directory entry.

### Search

Given a particular key k, we search for a record by first looking in the directory, **D**. If there is no directory entry for the key, the record is missing, represented thus:

$$\text{absent:} \quad k \ \textit{esearch} \ \mathbf{D} = \Lambda \tag{8.2.1}$$

If the record is present, *esearch* finds an entry containing the desired key and a pointer, so that we have

$$\text{present:} \quad k \ \textit{esearch} \ \mathbf{D} = \mathbf{D}_i \tag{8.2.2}$$

In Fig. 8.2.2 we perform a search for bob and find that the fourth directory entry has the key we are looking for:

$$\text{bob} \ \textit{esearch} \ \mathbf{D} = \mathbf{D}_4 \tag{8.2.3}$$

**Fig. 8.2.2.** Besides the list area **L** which may contain variable length records, there is a directory area, **D**, kept ordered so we can easily find a record.

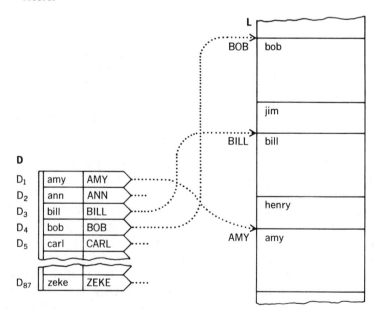

Having found an entry with the desired key, we extract its pointer. It is important to note that the position of the record in the file is generally not the same as the position of the entry in the directory, so that we have

$$point \text{ k } esearch \textbf{ D} = point \textbf{ D}_i = \textbf{L}_j \qquad (8.2.4)$$

Now we go to the location designated by the pointer and we should find the desired record. Symbolically we have

$$(point \text{ k } esearch \textbf{ D}) = (point \textbf{ D}_i)_\textbf{K} = (\textbf{L}_{j\textbf{K}}) = \text{k} \qquad (8.2.5)$$

With reference to the search in progress for bob, we have

$$(point \text{ bob } esearch \textbf{ D}) = (point \textbf{ D}_4)_\textbf{K} = (\textbf{BOB})_\textbf{K} = \text{bob}$$
$$(8.2.6)$$

### Appending

Figure 8.2.3 illustrates how a record is appended using the total directory system. Before appending the record, verify that it is absent by searching the directory:

$$(\text{ben } egsearch \textbf{ D}) = (\textbf{D}_3) = \text{bill} \qquad (8.2.7)$$

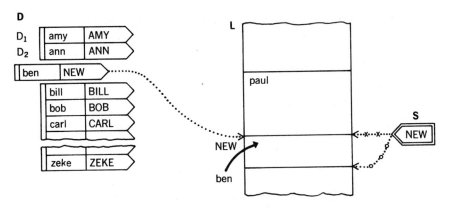

**Fig. 8.2.3.** Appending a record with key ben, and a new directory element for him

The search reveals that the record is absent and that when we create a directory entry for ben, it should precede the entry with key bill at $\textbf{D}_3$.

To keep track of space in the list, L, use the space pointer, S, as shown in the figure. This pointer now indicates that empty space is found in L at a place that we symbolically refer to as NEW.

The record, ben is placed in the list area at NEW:

$$\text{ben} \rightarrow point\ S = \text{NEW} \qquad (8.2.8)$$

The figure then shows the space pointer updated thereafter.

Now update the directory. The entry for ben should occupy the cell, $D_3$. All other entries must thereafter be moved down by one entry. Figure 8.2.3 shows the new entry containing the key ben and the pointer NEW entered into the directory, and we have

$$\text{ben} \rightarrow D_{3K}; \qquad \text{NEW} \rightarrow D_{3P} \qquad (8.2.9)$$

### Deletion

Deletion requires that a record currently in the file be eliminated. With the space pointer technique, the record need not be eliminated from $L$, but the cell that it occupies is no longer available. For the space list technique, the cell may be returned to the space list and reused.

In any case, the directory must be updated. There are two ways to do this, both of which reflect deletion for dense ordered lists:

- The entry may be removed by moving other entries upward, thus obliterating the deleted entry.
- The entry may be kept but a delete symbol ($\phi$) is inserted in the key position to indicate that the record is no longer available.

The latter method has one advantage and one disadvantage:

- The directory is quickly updated.
- The entry that reflects deletion may have to be inspected during many future searches, however.

### Characteristics

The directory is usually ordered and dense; then it is possible to do a quick serial search or, more preferably, a binary search of the directory. The list may be large but it is still easy to add new records to it. Variable length records are no problem.

On the other side of the ledger, it is time-consuming to append to and delete from the file since the directory must then be reorganized. If the file is stable, there is no problem. If it is subject to many changes, the linked list total directory system described in the next section might be preferable or, better still, the binary tree total directory of Section 11.1.

## 8.3 LINKED LIST DIRECTORY FILE

We have seen that when the total directory is kept as an ordered list, it is hard to alter. The linked list is easy to alter. The linked list directory is easy to alter also but can be searched only serially. The binary tree directory file of Chapter 11 has both advantages.

### File Structure

We actually encountered the linked list directory file in Chapter 6 where the linked list carries a data pointer instead of a datum. The linked list cell appears in Fig. 8.3.1, where cell D2 contains three parts:

- The cell contains an extra vertical line on the left to indicate that it is part of a directory.
- The key, $D2_K$, is the key of the record (ann).
- The link, $D2_P$, is a pointer to the next directory element, D3.
- The data pointer, $D2_D$, provides the location of the record (ANN).

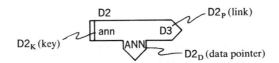

**Fig. 8.3.1.** The linked list directory element

Figure 8.3.2 shows a linked list directory file that starts with the directory head, **D**. The directory head points to the first cell in the directory at D1. The figure shows how cells of the directory are chained together. On the right of the figure the list area, **L**, contains records in any order and of any size, even variable. Data pointers in the directory take us to the corresponding records in **L**.

### Search

To find a record, given its key, first search the directory. For instance, to find **bob** we do the following:

$$\text{bob } \textit{esearch } \mathbf{D} = \mathbf{D4} \tag{8.3.1}$$

If we encounter a record with a higher key, the search does not stop since the linked list may not be arranged in order. For an ordered (by key) linked list the search can be made to stop using the operator *egsearch*.

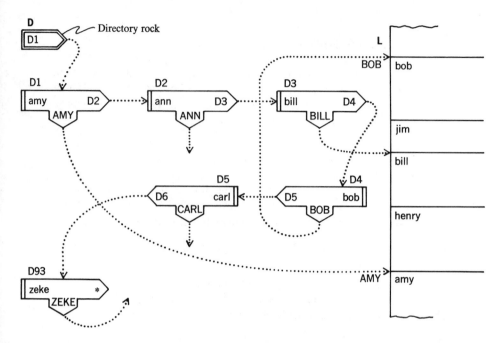

**Fig. 8.3.2.** Showing a linked list directory and the list to which it applies

Once we reach the record, we take out the data pointer with a pointer operator. If we use point, it will only take out the link. To take out the data pointer, let us propose another operator, *data*. Then to find the record, apply *data* to the cell and obtain the desired location:

$$data \text{ bob } esearch \textbf{ D} = data \text{ D4} = \text{BOB} \qquad (8.3.2)$$

Naturally the record we have found has the key originally provided. Thus:

$$(data \text{ bob } esearch \textbf{ D})_\text{K} = (\text{BOB}_\text{K}) = \text{bob} \qquad (8.3.3)$$

### Append

To alter a list requires two space pointers defined with reference to Fig. 8.3.3:

- The list space pointer **SL** points to available space for records in the list area **L** and in the figure points to NEW (1).
- The directory space pointer, **SD**, points to new cells for the directory, and in the figure the next available cell is D108 (2).

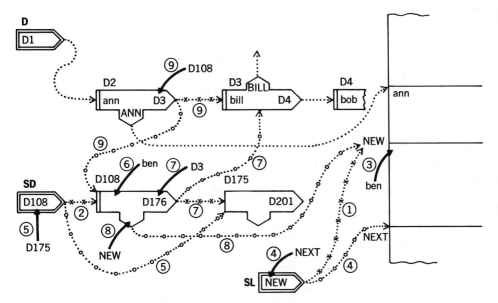

**Fig. 8.3.3.** Appending a new record, ben, and a new directory element for him

The directory space list is kept as a linked list but only the links are filled in.

To append a new record named ben, first search the directory and find that ben should be placed before D3:

$$\text{ben } egsearch \text{ } \mathbf{D} = \text{D3} \tag{8.3.4}$$

We determine that the new directory entry follows D2. Next, go to the space list head and *dequeue* space for our new record at NEW (1):

$$dequeue \text{ } \mathbf{SL} = \text{NEW} \tag{8.3.5}$$

ben is placed there (3); the pointer is updated (4):

$$\text{NEXT} \rightarrow \mathbf{SL} \tag{8.3.6}$$

Now, go to the space directory for a directory cell:

$$dequeue \text{ } \mathbf{SD} = \text{D108} \tag{8.3.7}$$

The directory space pointer is then updated (5). The record key is placed into the cell (6) and the cell is pointed to D3 (7). The pointer to the record NEW is inserted in the directory cell (8). Finally, the predecessor, D2, is linked to the new cell (9).

### Deletion

To delete a record, simply delete the directory entry. The deleted directory cell is returned to the directory space list and the predecessor directory entry has its link updated.

The record itself may remain untouched in the list area or it may receive a delete tag so that it will not be referenced later.

## 8.4 SINGLE DIRECTORY

To make a neighborhood directory or sublist directory for a list, the list should be

- ordered;
- dense.

The sublist directory file or simply the **directory file** assumes that the list (call it **A**) is divided into sublists. Let us say that there are n sublists. Each sublist could be different in size. To simplify explanation, assume the same number of cells in each sublist.

### Principle

The directory technique is really quite simple:

- Take an ordered dense list, **L**, and divide it into a number of sublists.
- Form a directory, **D**.
- For any sublist, $_iL$, make one directory entry containing
  * a starting point for lookup in the sublist, the sublist pointer, $D_{iP}$;
  * a key value to indicate the range of records in the sublist, the sublist key, $D_{iK}$.

For each entry we choose

- for the pointer, the address of the first record in the sublist;
- for the key, that of the last record in the sublist.

This provides the address at which to start looking. For the key range:

- The key in the *previous* directory entry sets a *lower* limit for records in this sublist.
- The key in *this* entry sets an *upper* limit for records in this sublist.

- The first sublist has no explicit lower bound; use the lowest possible record key to set the lower limit.

### Limitation

For the present, we examine only a dense list. It presents an inherent obstacle to appending and deleting. We postpone remedies until Chapter 10 on overflow. Now assume that the list is only referenced, not changed—we only *search* the directory list.

### Example

Figure 8.4.1 shows an example:

- The list starts at cell 300.
- Each cell contains five words (or bytes).
- Each sublist has six cells.
- The list holds thirty-six records.
- The directory starts at 600 and contains six one-word entries.

*search*  Let us look for the record with the key, jim. First we search, *equal greater*, the directory for an entry using the key, jim.

- Start with the first entry at 600.
- It is left justified (both the sublist key and the sublist pointer start at first position at the left) so that entry and key sought can be compared directly.
- Stop at 603 where joe > jim.

We have pinpointed the sublist—it starts at 390.

- Extract the sublist pointer from the directory by masking (actually the *point* operator does just this).
- Do a *search equal* for jim starting at 390.

It is clear that we start looking at irv and find jim on the fifth look at the sublist.

*record absent*  Let us look for paul. *We* can see right away that he is not there. It takes a little longer for the COMPUTER to find this out (in number of operations, not time, of course).

- The directory shows that we want the sublist where phylis lives.

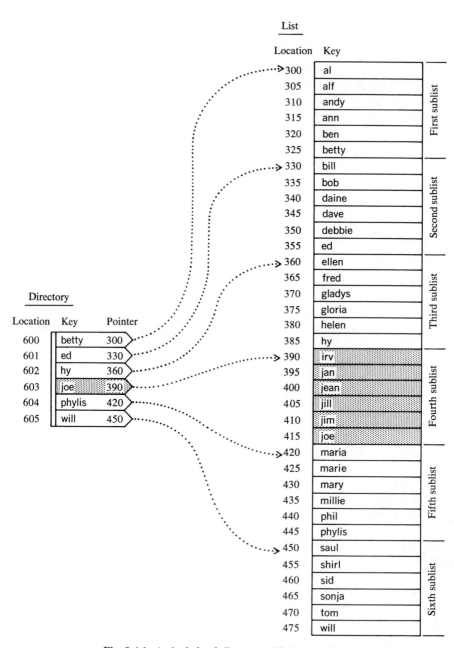

**Fig. 8.4.1.** A single level directory with the six sublists to which it applies

- We start looking at maria.
- When we reach phil, we know paul is not there since paul < phil.

### Number of Looks

One criterion for comparing list structures is the number of looks it takes to find an item or verify that it is missing. For the directory method this is simple to calculate:

- On the average, we look halfway through the directory.
- On the average, we look halfway through the sublist after the directory finds the right sublist for us.

Let us compare directory look up with other methods—binary and linear search. To be fair, let us take a good-sized list—say 10,000 records (N = 10,000)—and assume 100 sublists of 100 records (n = 100). Then we have the following:

- Linear search requires $N/2 = 5000$ looks.
- Directory search requires $n/2 + n/2 = 100$ looks.
- Binary search requires about $\log_2 (N - 1) = 13$ looks.

Looks like we have struck a happy medium and we don't have all those fence calculations that binary search requires!

### Optimum Sublist

We placed an arbitrary number of records into each sublist. What is the best allocation? Suppose there are N records in the list and we place X records in each sublist. Then there are N/X entries in the directory. The number of looks, L, is given as

$$L = \frac{1}{2}\left(\frac{N}{X} + X\right) \qquad (8.4.1)$$

The optimum value is found by differentiating both sides with respect to X:

$$0 = \frac{N}{-2X^2} + \frac{1}{2} \qquad (8.4.2)$$

Transposing and multiplying by $2X^2$, we have

$$X^2 = N$$
$$X = \sqrt{N} \qquad (8.4.3)$$

The best distribution results when we have $\sqrt{N}$ sublists each with $\sqrt{N}$ cells. Of course that is what we used above in the example of 10,000 records.

### Specification

Consider our list, **L**. We now know that it is best to divide it into n sublists where

$$n \doteq \sqrt{num\ \mathbf{L}} \tag{8.4.4}$$

Call a general sublist, the ith, $_i\mathbf{L}$; then the list, **L**, is the collection of these sublists:

$$\mathbf{L} = \{_i\mathbf{L}\}; \quad (i = 1 \text{ to } n) \tag{8.4.5}$$

And there is a directory, **D**, composed of n entries:

$$\mathbf{D} = \{D_i\}; \quad (i = 1 \text{ to } n) \tag{8.4.6}$$

A directory search is written as

$$k\ egsearch\ \mathbf{D} = D_j \tag{8.4.7}$$

and the pointer is extracted with

$$point\ D_j = {}_j\mathbf{A} \tag{8.4.8}$$

The sublist is searched thus:

$$k\ esearch\ {}_j\mathbf{A} = {}_j\mathbf{A}_i \tag{8.4.9}$$

where the record is found. Put this all together and we find

$$(k\ esearch\ point\ egsearch\ \mathbf{D})_K = (_jA_{iK}) = k \tag{8.4.10}$$

## 8.5 TWO-LEVEL DIRECTORY FILE

We have seen that $\sqrt{N}$ is an optimum number of cells per sublist for a simple directory file. Another alternative is to keep the sublists small and let the directory grow large. Now, instead of using the directory directly, create another directory for expediting the use of the original directory. This is becoming confusing, isn't it? Let us think up better names.

Divide the original directory into subdirectories. Then we have

- a master directory, which leads to one of several subdirectories;

- a subdirectory, which leads us to one of several subsublists—a subdivision or fraction of a sublist;
- the master directory entry corresponds to one sublist consisting of several subsublists.

**Example**

An example should clear the air. In Fig. 8.5.1 we have

- a directory with six entries;
- six subdirectories of six entries each;
- a list of 216 cells with room for five words (bytes) each—this is divided further:
  * six sublists of thirty-six cells each;
  * thirty-six subsublists of six cells each.

Notice the makeup of each directory entry:

- the key of the last record in the sublist;
- a pointer to the first entry in the subdirectory for that sublist.

*search*    Let us try to find pepe:

1. Search the directory for *equal greater* to obtain robin.
2. Get the subdirectory address for robin, 624.
3. Search the robin subdirectory for *equal greater* to obtain pisano.
4. Get the subsublist address for pisano, 1910.
5. Search the pisano subsublist for *equal greater* to obtain pepe.

The number of looks is $5 + 4 + 3 = 12$; the average number of looks is 9.

*record absent*    Let us look for robert, who is missing. We proceed as in the section above:

1. Search for robert in the directory to obtain robin, 624.
2. Go to the subdirectory.
3. Search for robert and find robin there too.
4. Go to 1970, the beginning of the subsublist.
5. At 1995, find robin > robert; hence robert is absent.

**Number of Looks**

It can be shown that the optimum size for the subsublist is $\sqrt[3]{N}$ so that the following are all about equal in size:

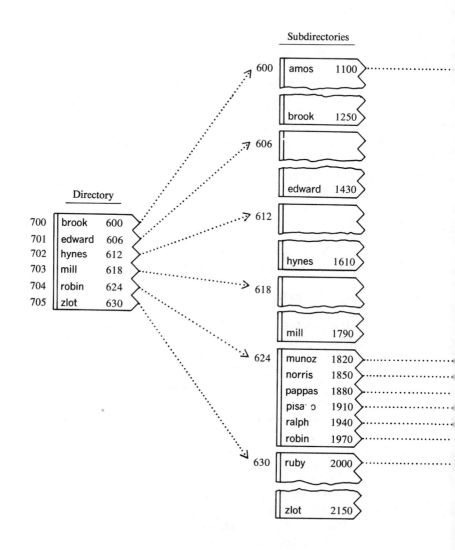

**Fig. 8.5.1.** A two-level directory illustrating searches for both present and absent records

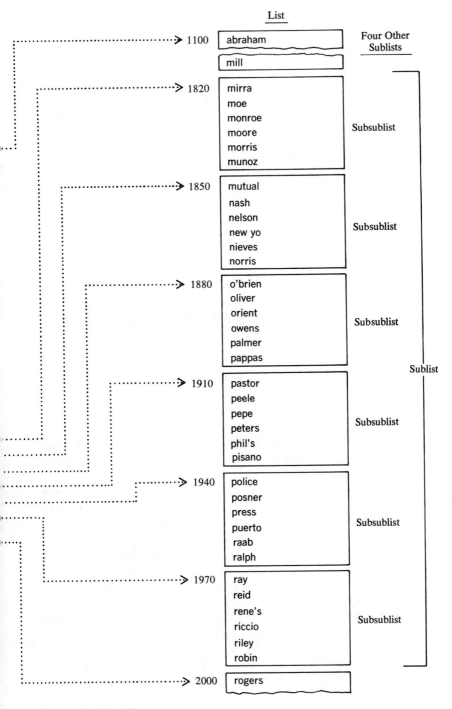

**Fig. 8.5.1.** (*Cont'd.*)

- the directory
- the subdirectory
- the subsublist

If each is equal to n in size, the average number of looks for a search is the same. We examine half of the subsublist, the subdirectory, and the directory where each has n cells.

To find some idea of the effectiveness of the search, let us make a comparison for 1000 records:

- serial search:      $L = N/2 = 500$;
- directory search:   $L = n = \sqrt{N} = 31.6$;
- double directory:   $L = 3n/2 = 3\sqrt[3]{N}/2 = 15$;
- binary search:      $L = [\log_2 N] = 10$.

### General Case

We have seen a specific example in Fig. 8.5.1 and now we address ourselves to the general case of a two-level directory referring to Fig. 8.5.2.

*subsublist*   The first thing to be done is to calculate n, given by
*size*

$$n = \sqrt[3]{N} \qquad (8.5.1)$$

If the size of our list N is 125, the directory size n is the cube root of N and $n = 5$ (8.5.1). This key number also provides the size of each subdirectory and each subsublist.

One of the laws of COMPUTER lore is that nothing ever works out exactly. Eventually we shall be asked to partition, say, a list of 100 records. When we take the cube root, we obtain some oddball answer. Actually we could use n of 5 and have some extra room in our file. If we do not have a good means for adding records, however, the room will go to waste. Another alternative would be to set our sublist size equal to 4 and our subdirectory size equal to 5. This would work out to be exactly 100. Even this is too good to be true. What do we do to set up the list for 107 records?

Thus (8.5.1) is a guide and the actual design parameters may be determined empirically from it. Suppose now that we have derived n.

*sublists*   Divide **L** into n sublists, distinguishing each of these by a left subscript, so that we have

$$\mathbf{L} = \{_i\mathbf{L}\} \qquad (i = 1 \text{ to } n) \qquad (8.5.2)$$

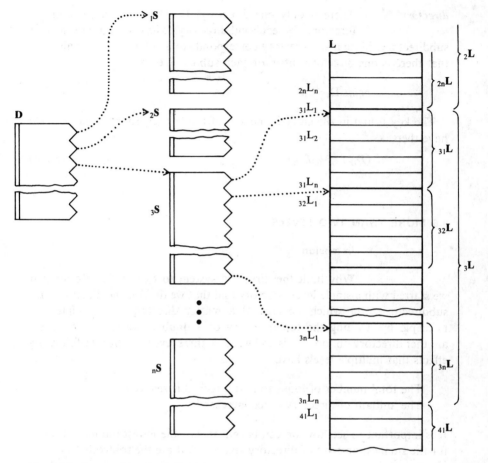

**Fig. 8.5.2.** A general two-level directory

Each sublist is further divided into n subsublists. Thus:

$$_iL = \{_{ij}L\} \qquad (j = 1 \text{ to } n) \tag{8.5.3}$$

**subdirectory**     There is one subdirectory S for each sublist. Use a left
                    subscript to designate the subdirectory. Thus for the third
sublist $_3L$, use the subdirectory $_3S$. Each entry in the subdirectory points to
the beginning of a subsublist of the sublist to which it belongs. The five
subsublists to which $_3S$ points are shown clearly in Fig. 8.5.2.

   The key portion of the entry in a subdirectory is the key of the *last* record
of the subsublist to which the entry points or, symbolically,

$$(_{ij}S_K) = (_{ij}L_{nK}) \tag{8.5.4}$$

**directory**    There is only one directory, **D**. There is one entry in the directory for each subdirectory. The entry points to one subdirectory. Since a subdirectory corresponds to a sublist, we may also say that there is one directory entry for each sublist. We have

$$point\ D_i =\ _iS \tag{8.5.5}$$

The key found in the entry is the key of the last record in the sublist. We have then

$$(D_{iK}) = (_{ni}L_{nK}) \tag{8.5.6}$$

## 8.6 MORE THAN TWO LEVELS

### Extension

Why limit the directory system to two levels? Recall that we started with a single-level directory and that we divided the directory into subdirectories for which we created a master directory. Extrapolate this principle by dividing the top directory of a double directory system. Set another directory on top of this and we have three levels. Notice the following effects that multiple levels have

- The total number of looks to find a record is reduced.
- The amount of directory space increases.

It is important to see how we can benefit most—the fewest number of looks for the smallest amount of directory space. What are the tradeoffs?

### Four Levels

To see the directory principle extended further, Fig. 8.6.1 shows a system with four levels of directories sitting on top of the list. We have picked a perfect size for the list; it consists of 3125 records, which is the fifth power of 5. We have then

$$N = num\ L = 5^5 = 3125; \qquad n = \sqrt[5]{3125} = 5 \tag{8.6.1}$$

The list consists of five sublists, each of which contains 625 records. The top directory contains five entries. We have then

$$D = \{D_1, D_2, D_3, D_4, D_5\} \tag{8.6.2}$$

$$L = \{_1L,\ _2L,\ _3L,\ _4L,\ _5L\} \tag{8.6.3}$$

$$num\ _iL = 625 \tag{8.6.4}$$

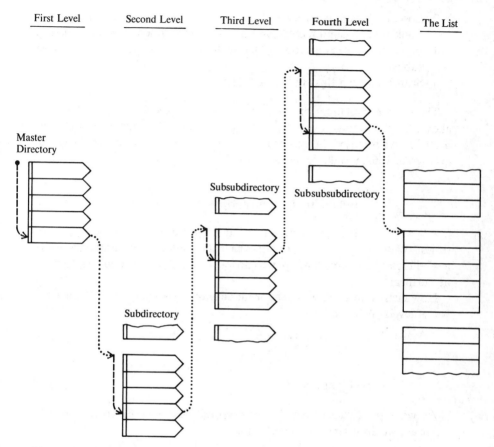

**Fig. 8.6.1.** A four-level directory system

The first part of a directory entry is the key of the last record of the sublist to which it refers. The pointer part of the entry is to the subdirectory on the next lower level for the subsublists that make up *that* list. There are five second-level subdirectories and each consists of five entries. That describes level 2.

Each subdirectory at level 2 contains five entries, each of which points to a subsubdirectory on level 3. Thus there are twenty-five subsubdirectories on level 3; each applies to a subsublist! By the same token, each entry in a subsubdirectory on level 3 points to a subsubsubdirectory on level 4, of which there are consequently 125. Thus there are 625 total entries on level 4. Each points to a low level sublist consisting of five records.

*search*          The procedure to find a record should be apparent:

1. Look in the directory for a pointer to a subdirectory.

2. Look in the subdirectory and find a pointer to a subsubdirectory.
3. Look in a subsubdirectory and find a pointer to a subsubsubdirectory.
4. Look in the subsubsubdirectory to find a pointer to the lowest level sublist.
5. Search through this sublist.

There are a maximum of five items to be examined for each of the steps above. On the average we go through half of them. Therefore the total number of looks required to find our record is 12.5; this is contrasted with an average of 1562.5 looks for serial search.

### Twin Entry Directory System

How far can we continue to pile directories on top of directories? Well, we shall have to stop when the top directory and each lower level directory contains exactly two entries; it would do no good to split them up further!

How many levels will we have for the **twin entry directory**? Call T the number of levels; then we have

$$2^{T-1} \leq N < 2^T \qquad (8.6.5)$$

or

$$T = \lceil \log_2 N \rceil \qquad (8.6.6)$$

As we expect, these formulas correspond closely to the binary search. After all, we are doing the same action:

- The top directory determines which half of the list to look in.
- The next level tells us which quarter.
- The next level tells us which eighth, and so forth.

Figure 8.6.2 shows diagrammatically how the twin entry directory works.

### Contrast

Of what avail is it to pile on more directories? Two tables are provided to show the contrast. Table 8.6.1 compares various numbers of levels of directory files for a file size of 10,000. Table 8.6.2 does the same thing for a file size of 1 million. Keep in mind that as the directory space mounts, it still does not come close to the file size since the directory entry is much smaller than the record itself.

The single-level directory provides a dramatic improvement over the linear

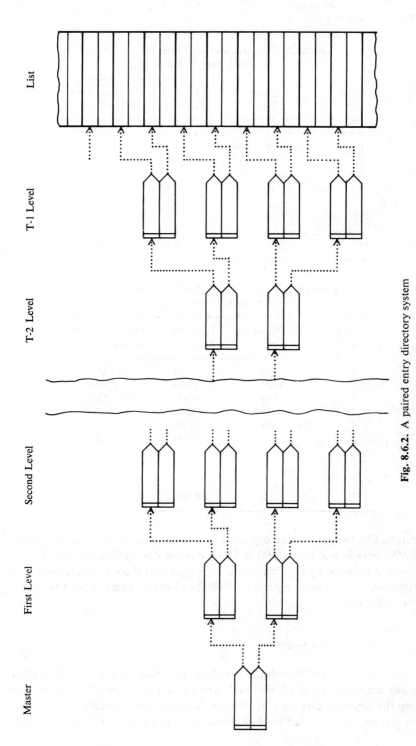

**Fig. 8.6.2.** A paired entry directory system

235

**Table 8.6.1.** Comparison of Multilevel Directory Files
for a File Size of 10,000

| Number of Levels | Directory Size | Directory Space | Average Number of Looks |
|---|---|---|---|
| 0 (total) | 0 | 0 | 5000 |
| 1 | $\sqrt{N} = 100$ | 100 | 100 |
| 2 | $\sqrt[3]{N} = 21.54$ | 486 | 32 |
| 3 | $\sqrt[4]{N} = 10$ | 1110 | 20 |
| 4 | $\sqrt[5]{N} = 6.31$ | 1883 | 16 |
| $\vdots$ | | | |
| 13 | $\sqrt[14]{N} = 2$ | 9998 | 14 |

**Table 8.6.2.** Comparison of Multilevel Directory Files
for a File Size of One Million

| Number of Levels | Directory Size | Directory Space | Average Number of Looks |
|---|---|---|---|
| 0 | 0 | 0 | 500,000 |
| 1 | 1000 | 1,000 | 1,000 |
| 2 | 100 | 10,000 | 150 |
| 3 | 31.6 | 33,000 | 63 |
| 4 | 16 | 66,000 | 40 |
| 5 | 10 | 100,000 | 30 |
| $\vdots$ | | | |
| 20 | 2 | 999,998 | 20 |

search. The two-level directory takes us from 100 to 32 looks in the case of 10,000 records and from 1000 to 150 looks for one million records. Both of these are noteworthy improvements. As we go past three levels, however, the improvement becomes very small, while the directory space required increases tremendously.

### Separation

We have shown the directories and the main list all together in our diagrams. A good approach when files are on auxiliary media is to keep the top level directory in MEMORY. The next level directory is found on the volume, and another level of directory kept possibly with a subdividion of the volume, such as the track or the disk pack. This keeps the amount of space occupied in MEMORY at a minimum, reduces the number of accesses, but maintains the effectiveness of the system.

## PROBLEMS

8.1  Contrast the meaning of *neighborhood, sublist, block,* and *subfile.*

8.2  For neighborhood locatability, what are the alternatives to
(a) determine the neighborhood in which to look?
(b) perform the search or probe?

8.3  What are two conditions, either of which might suggest use of a *total* directory?

8.4  A directory $\mathbf{D}$ contains some entry $D_i$ with a key position $D_{iK}$ and a pointer position $D_{iP}$. From this, how do we determine
(a) the starting and ending address of the neighborhood, $L_i$?
(b) the least and greatest key of records in $L_i$?

8.5  Explain in *words* for the total directory:
(a) finding a record given its key;
  · record present
  · record absent
(b) appending a new record for the list; the directory;
(c) deleting an old record from the list; the directory.

8.6  Draw a *colored* directed tree to represent the total linked list directory file (TLLDF) using *different* colors for these relations for the directory $\mathbf{D}$ and the list $\mathbf{L}$:
(a) successor;
(b) key greater than;
(c) next entry pointer, directory;
(d) data pointer.

8.7  Repeat Problem 8.5 for the TLLDF.

8.8  For the single-level directory file (SLDF):
(a) How is the list of records, $\mathbf{L}$, organized?
(b) How is the directory, $\mathbf{D}$ organized?
(c) What is in the directory entry?
(d) Describe the size in bytes of $\mathbf{L}$ and $\mathbf{D}$.

8.9  Describe in *words* for the SLDF how to search for a record that is present; absent.

8.10  For the two-level directory file (TLDF):
(a) What determines the entry size for
  · the list $\mathbf{L}$?

- the subdirectory S?
- the directory D?

(b) What determines the *size* (number of entries) for **L**, **S**, and **D**?

(c) How is each of **L**, **S**, and **D** organized?

8.11 We have 200,000 records of 500 bytes each in our file. Each is identified by an 8-byte key. Assume a directory entry requires 12 bytes. Consider the following alternatives:

serial list    three-level directory file

SLDF        four-level directory file

TLDF        five-level directory file

(a) For each of the above, how much storage is required?

(b) For each of the above, what is the size (number of records) of the list, sublist, subsublist, etc.?

(c) What is the size (number of entries) of the directory, subdirectory, etc.?

(d) What is the average number of looks required to find a record for the six lists?

8.12 What is it about the multilevel directory file that reduces the number of looks required?

8.13 For the twin entry directory file (TEDF) of *n* records:

(a) How many levels are there?

(b) How many entries are there at each level?

(c) How many looks does a search entail?

(d) For record size, *rs*, and entry size, *es*, what is the format for the space required?

8.14 Do Problem 8.11 for the TEDF.

8.15 Make an SLDF called **f1** out of the list in Table 8.P.1 which gives the location of each single field record.

(a) Select an optimum sublist and directory size.

(b) Choose an entry size for the directory.

(c) Start the directory at location 7000 opening entry according to (b).

(d) Show the contents of each directory entry.

8.16 For **f1**, show how to look for

(a) Fine,

(b) Fink,

(c) Lamb,

(d) Manzella.

8.17 For Table 8.P.1, store the file on a 3330-disk pack starting in cylinder 100 head 0. Block the 100-byte records so that each block contains one sublist. Now show the directory.

8.18 Make a TLDF called **f2** out of Table 8.P.1.

(a) Choose the size of sublist, subsublist, directory, and subdirectory and the entry size.

(b) Start the subdirectories at 1000 and put the directory after it.

(c) Show the subdirectories and directory.

8.19 For **f2**, show how to look for

(a) Hill,

(b) Hall,

(c) Kole,

(d) Kolb.

8.20 Make a TEDF, **f3**, out of Table 8.P.1 where the top level directory starts at 9000, the next level subdirectories start right after, the next level after that, and so forth.

8.21 For **f3**, show how to look for

(a) Hill,

(b) Hall,

(c) Kole,

(d) Kolb.

8.22 Create an SLDF, **v**, for Table 8.P.1, consisting of **V** and **D$_v$**. **V** is a list where each record contains a *name*, a key field of *variable* length, and a *telephone number* of seven characters and [**V**] = 55,000. There are thirteen sublists, twelve containing sixteen records. Each directory entry uses four characters for a truncated key and five for a pointer. Sketch **V** and **D$_v$**.

8.23 For **v**, show how to find

(a) Hill,          (e) Lamb,

(b) Fink,          (f) Kolb,

(c) Kole,          (g) Hall,

(d) Manzella,      (h) Fine.

8.24 Create a TLDF, **u**, using **U** with records of variable length as in Problem 8.22 and directory **D$_u$** and subdirectories **S$_u$** with entries as in Problem 8.22 where

$$[U] = 70,000; \quad num\ U_1 = 36$$
$$[S_u] = [U] + lin\ U; \quad num\ S_{u1} = 6; \quad num\ S_{u6} = 4$$
$$[D_u] = [S_u] + lin\ S_u; \quad num\ D_u = 6$$

Sketch.

8.25 Repeat Problem 8.23 for **u**.

8.26 Create a TEDF, **w**, using **W** as described for **U** above. The lowest level subdirectory follows **W** and then the next lowest and so forth until the final twin entry directory. Each entry is nine bytes as above. Sketch.

## Table 8.P.1

| | | | |
|---|---|---|---|
| 34600 | DeAngelo Salvatore | 39600 | Frankel Tobias |
| 34700 | DeFalco F | 39700 | Frater Basil |
| 34800 | Delany Miles V | 39800 | Freese Deborah |
| 34900 | Delson Richard | 39900 | Friedlander Alan P |
| 35000 | Demetropoulos Alex | 40000 | Friend Howard L |
| 35100 | DePalma Ralph | 40100 | Fry Edw Donald |
| 35200 | Desiderio Patrick | 40200 | Furstman Dorothy |
| 35300 | DeVito Michl F | 40300 | Gagliardi Luciano |
| 35400 | DiBlase Frank | 40400 | Gallant H |
| 35500 | Di Fazio Elec Inc | 40500 | Gannon Martin F |
| 35600 | DiMattia Eddie | 40600 | Garey Joel L |
| 35700 | Disend Leo | 40700 | Gasparini Alfiero |
| 35800 | Dodici Richard | 40800 | Gehrig S |
| 35900 | Donaghy John J | 40900 | Genco A |
| 36000 | Donovan Thos J | 41000 | Gentile Louis |
| 36100 | Dougan Robt A | 41100 | German D |
| 36200 | Dozier Dollie | 41200 | Giamo Joseph |
| 36300 | Droutman Gary | 41300 | Giglio CC |
| 36400 | Duffy Michael | 41400 | Gilmartin Jason C |
| 36500 | Dunne John J | 41500 | Giovitto Patsy |
| 36600 | Dykas Toni | 41600 | Glass Valentine L |
| 36700 | Eastburn John | 41700 | Glenwick Henry |
| 36800 | Econ Olga Mrs | 41800 | Goehler Karl |
| 36900 | Edwards Robert W | 41900 | Goldberg Jack |
| 37000 | Einhorn Frederick | 42000 | Goldin Bert |
| 37100 | Electromatic Equipt Co | 42100 | Goldstein Edw |
| 37200 | Elmont Cemetery Inc | 42200 | Gomes John D |
| 37300 | Endo Lester | 42300 | Goosten Ronald |
| 37400 | Epstein Charles | 42400 | Gorman John D |
| 37500 | Erwin Hugh | 42500 | Govoni A F |
| 37600 | Evangelista Anthony | 42600 | Grand Manny |
| 37700 | Faber A | 42700 | Gray Joseph H |
| 37800 | Falco Thos | 42800 | Greco Marie A |
| 37900 | Farella Frank | 42900 | Greenberg David |
| 38000 | Fasini Thos | 43000 | Greenfield Seymour |
| 38100 | Fedrow Benj | 43100 | Grella Philip |
| 38200 | Feldis J | 43200 | Grinberg Bessie Mrs |
| 38300 | Fenimore Anthony R | 43300 | Grosser Allen |
| 38400 | Ferrari John P | 43400 | Gruskoff Paul |
| 38500 | Fick Francis H | 43500 | Guille John C |
| 38600 | Fine L | 43600 | Gutbrod Wm G |
| 38700 | Fiore Guido | 43700 | Haber Harold |
| 38800 | Fischetto A | 43800 | Haight Gary |
| 38900 | Fitzgerald Jos | 43900 | Halloran Richd |
| 39000 | Flapan Robert M | 44000 | Hammond Richd J |
| 39100 | Floral Park Florists | 44100 | Hansell Harry |
| 39200 | Foisy Emilie C | 44200 | Harloff C S |
| 39300 | Forest Helen Mrs | 44300 | Harris Seymour |
| 39400 | Foster Christopher J | 44400 | Hartwell Helen |
| 39500 | Fraenkel Jerome J | 44500 | Hauer Irving |

**Table 8.P.1.  (Cont.)**

| | | | |
|---|---|---|---|
| 44600 | Hazel Robert B | 49600 | Koconas Christopher |
| 44700 | Herrernan Neil | 49700 | Kolb Jos A |
| 44800 | Helfgott Harry | 49800 | Kopivsek Jerry |
| 44900 | Hempton Francis | 49900 | Kotarski John C |
| 45000 | Henle Andrew | 50000 | Kramer Carolyn |
| 45100 | Herman Bernard | 50100 | Krauthamer Marilyn |
| 45200 | Hertz Reuben | 50200 | Kritzer Jesse |
| 45300 | Hickey John F | 50300 | Kubacki Henry |
| 45400 | Hill James M | 50400 | Kuraner Henry D |
| 45500 | Hirsch Howard | 50500 | LaBagh Victor A |
| 45600 | Hockenjos Ruth Mrs | 50600 | LaFlare Mary J |
| 45700 | Hoffman Ronald | 50700 | Lamberson Hugh H |
| 45800 | Holland Edw | 50800 | Landry Francis W |
| 45900 | Holzberg Rose Mrs | 50900 | Lanig Jos |
| 46000 | Horan Donald J | 51000 | LaRose Robert W |
| 46100 | Hoskinson Lara Joan | 51100 | Lauda Francis C |
| 46200 | Howie Louise A | 51200 | Lawrence Lee |
| 46300 | Hulings Wade | 51300 | Leas Frank J |
| 46400 | Hurwitz Amy | 51400 | Lee Thomas |
| 46500 | Iaboni August | 51500 | Leibowitz Gerald |
| 46600 | Imperial John J | 51600 | Lennox Leo |
| 46700 | Intravia Jas | 51700 | Lerner Howard |
| 46800 | Irwin J | 51800 | Leventhal Lewis |
| 46900 | Iuliucci Felix | 51900 | Levine Melvin |
| 47000 | Jackson Abraham | 52000 | Levy David |
| 47100 | Jacobson Edwin | 52100 | Lewis John M |
| 47200 | James Lillian Mrs | 52200 | Lichtenwalf L |
| 47300 | Jay B | 52300 | Liguori Jas H |
| 47400 | Jeser Morris | 52400 | Lindros Everett G |
| 47500 | Johndro M | 52500 | Lipton Morris J |
| 47600 | Johnson Lucille | 52600 | Lobdell Jas G |
| 47700 | Jones Geo | 52700 | Logan Walter J |
| 47800 | Joy John L | 52800 | Long Joseph A |
| 47900 | Kadane David K | 52900 | LoRusso Michael |
| 48000 | Kalkstein Benj | 53000 | Lowenthal Gerald |
| 48100 | Kanner Joshua H | 53100 | Luechau Linda |
| 48200 | Kapner Diane | 53200 | Lynch Chas F |
| 48300 | Kaslow Marilyn | 53300 | Lyons Clarence |
| 48400 | Katz Robt | 53400 | Macek Alexander |
| 48500 | Kaye Leo | 53500 | Madden Jesse |
| 48600 | Keeler S | 53600 | Magilligan Lawrence |
| 48700 | Kelly Edw | 53700 | Maidhof Geo |
| 48800 | Kempf Jacklyn Mrs | 53800 | Malin Herbert |
| 48900 | Keppler Arnold | 53900 | Manchester D |
| 49000 | Ketcham Francis | 54000 | Manley Wm J |
| 49100 | Kimmins Boyd | 54100 | Manzella Louis |
| 49200 | Kipling John | 54200 | March Terry |
| 49300 | Klamson Nick | 54300 | Margulies John A |
| 49400 | Klein & Teicholz | 54400 | Markowitz Meryl S |
| 49500 | Knight Theodore | 54500 | Marsden John M |

8.27  Repeat Problem 8.23 for **w**.

8.28  Contrast **v**, **u**, and **w** and **f1** and **f2** as to
   (a) space occupied,
   (b) number of looks to find present or absent items,
   (c) complexity to set up,
   (d) applicability to DASD.

# CHAPTER 9

# Mapping

## 9.1 INTRODUCTION

We have seen how the directory file enables us to find a neighborhood or sublist to start a serial search for a sought record. The directory itself is the means to localize the sublist; it, in turn, is used by a serial search technique. Is there an alternative?

**Mapping** is the alternative discussed here; it is an operation, arithmetic or otherwise, upon the sought key to yield the address of the **target sublist** or neighborhood. Search continues in this sublist using known techniques.

### Conversion

Consider a file **f** consisting of records each with a key to identify it. Mapping requires a reasonable set of keys—reasonable in that they lend themselves to some algorithm. The file **f** resides in a list **L**. The algorithm maps any record, $f_i$, into the cell occupied by $f_i$ in the list **L**, namely $[f_i]$. Use the operator, *map*, to represent this algorithm; then ideally we have

$$map\ f_i = [f_i] = L_j \tag{9.1.1}$$

where $L_j$ is the address of the cell supposedly occupied by $f_i$.

In certain select cases it is possible to implement such a mapping. Consider a small library. Books are assigned acquisition numbers according to their order of arrival at the library. A file is created describing these books, identifying them by acquisition number. One block on a DASD list holds each record. The relative block number *is* the address of the record *and* it *is* the acquisition number too.

A cross index might list books by topics. To retrieve a description, simply supply its acquisition number and that is its address in the DASD list:

$$map\ f_i = [f_{iK}] \qquad (9.1.2)$$

***counter-***         Consider the file in Table 8.P.1. The keys for this file are
***example***        names. If they were assigned successive cells in a list, one
                    would be hard put to find a "reasonable" algorithm to
convert a key into a relative cell location. Impossible, perhaps!

### A Hit-or-Miss Proposition

The techniques in this chapter use the key of a record to determine directly the place in MEMORY where the record is stored for retrieval or deletion or where it *should* be stored for appending. If we find the desired location immediately at the cell where we enter the neighborhood, this is a **hit**, which is

- the desired record for retrieval or deletion;
- an empty cell for append.

We are not so optimistic; we settle for a neighborhood that is then further examined using a probe, discussed later. A probe is done only if the first look is a **miss**, which is defined as

- the wrong record for retrieval or deletion;
- an occupied cell on an append.

Mapping turns up a cell that *contains* the desired record or delineates a sublist to start a search.

### Uniqueness

In the books-for-the-library example, the mapping is 1:1—any acquisition number uniquely determines the cell position. This is more than is required. We are satisfied to get a sublist location. This sublist contains several cells, one of which is the **target** cell, the cell that contains the record of interest. Also, since several keys are represented in the sublist, we expect *map* to take each of them into this same sublist.

### Record Population

To deal effectively with any **legitimate record**, one with a **legitimate key**, there should be rules for determining legitimate keys. For instance, a sought key of fifteen characters for a ten-character key field is obviously not legitimate. There may be limitations on the characters, so that in some cases only numerals are used; in others, only letters are used; in still others, letters and numerals may be used, but certain symbols such as punctuation marks may not appear.

For explanation, imagine the legal population, or universe, **p**, consisting of keys represented in Fig. 9.1.1. Also we have a list **L** containing records for

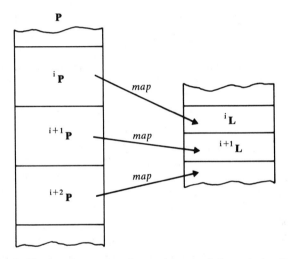

**Fig. 9.1.1.** The function $\chi$ maps from a key population **p** to locations in a list **L**.

*some* of the keys in **p**. Then a proper transformation (or mapping) operates on a legal key to produce a neighborhood or sublist, $_i\mathbf{L}$ of **L**. Define the operator, *map*, to do the mapping and we have

$$map \; \mathrm{p} = {}_i\mathbf{L} \tag{9.1.3}$$

where the subscript i denotes the ith neighborhood of **L** and p is some key. Here *map* transforms a key p into a location in **L** that is the start of $_i\mathbf{L}$. Actually we obtain the first cell in $_i\mathbf{L}$ called $_i\mathbf{L}_1$.

To be effective, a whole group of records from **p**, namely $_i\mathbf{p}$, should go to the same neighborhood:

$$map: \quad {}_i\mathbf{p} \rightarrow {}_i\mathbf{L} \tag{9.1.4}$$

where, generally,

$$map \; {}_i\mathbf{p} = {}_i L_1 \tag{9.1.5}$$

and ${}_i L$ is the sublist of $\mathbf{L}$, the neighborhood in which to search; ${}_i L_1$ is the first cell at which to look.

### Mapping Qualities

There are several properties that useful mappings may or may not have.

***order***    The key population, $\mathbf{p}$, presented in Fig. 9.1.1 is in order by key to make the presentation easy to comprehend. The list in MEMORY, $\mathbf{L}$, may or may not be in order by key—the mapping shown in the figure seems to preserve order. A mapping may be useful even though it is *not* order-preserving and we shall study some of these mappings later.

***many-to-one***    Useful mappings are **many-to-one**; that is, for any given key, p, there is exactly one neighborhood in the list to which the key applies—only one place that we need to search. There are many keys, however, that lead us to the same neighborhood. Hence, although the neighborhood defined for a key is unique, many keys translate to the same neighborhood address. A one-to-one mapping is impractical because it requires a list in MEMORY as large as the population.

***overlap***    In Fig. 9.1.2 we see neighborhoods resulting from an ordered mapping. Neighborhoods or sublists ${}_i L$ and ${}_j L$ are said to overlap when

$$\text{overlap:} \quad {}_i L \cap {}_j L \neq \Lambda \tag{9.1.6}$$

There is at least one cell in both sublists.

The list and its neighborhoods may be described as falling into three classes:

A. $\mathbf{L}$ may be nondense and the neighborhoods nonoverlapping and with space between them.
B. $\mathbf{L}$ may be dense or loose and the neighborhoods nonoverlapping but contiguous.
C. $\mathbf{L}$ may be dense or loose and the neighborhoods may be overlapping.

For (C) there is a danger that record order may become reversed where much revision occurs. We leave this to the reader to investigate.

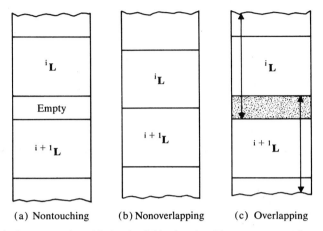

(a) Nontouching    (b) Nonoverlapping    (c) Overlapping

**Fig. 9.1.2.** Three kinds of neighborhoods with respect to overlap

*well-defined*    We say that *map* is **well-defined** when *map* p (with, perhaps, some implied parameter) completely determines the boundaries of the sublist $L_i$ where we look for p. Thus the directory method is generally well-defined since the sublist has exact boundaries; for the random probe we see later a search can encompass the entire list and hence is not well-defined.

## 9.2 ARITHMETIC MAPPING

The *map* operator may perform *calculations* on the key to determine the cell at which a neighborhood begins. We have these kinds of calculations:

- truncation,
- arithmetic,
- logic.

### Truncation

For simple **suffix truncation**, we might

- remove terminal alphabetic or alphanumeric symbols (suffix) from the field;
- remove the final digits for a decimal field;
- remove terminal bits from numeric fields that are
  * octal

* binary
* hexadecimal

For a numeric field, suffix truncation is equivalent to dividing by a power of the base. Thus, truncating a decimal field is division by a power of 10. Truncating a binary field divides by a power of 2, etc.

***example***    Figure 9.2.1 shows an example of address calculation by simple suffix truncation of a decimal key. The population, **p**, consists of 10,000 keys, each designated by a four-digit decimal number.

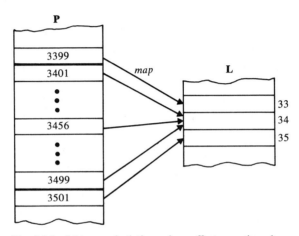

**Fig. 9.2.1.** Address calculation using suffix truncation above

Suppose the list **L** has 100 cells. There is one cell into which each of 100 keys in the population is mapped. In the figure, 34 is the address of the cell assigned to all records with key 34XX (where X is any digit). Thus, 3400, 3456, and 3499 all have the target address of 34. Then key 3399 maps to cell 33; keys 3500 and 3505 map to cell 35, etc. For this mapping $map_T$ we have

$$map_T\ p = \lfloor p/100 \rfloor \tag{9.2.1}$$

where the symbols $\lfloor \quad \rfloor$ mean to round down the number enclosed. Then for an example we have

$$map_T\ 3456 = \lfloor 3456/100 \rfloor = \lfloor 34.56 \rfloor = 34 \tag{9.2.2}$$

### Truncation and Relocation

If only truncation were done as described above, the list area would always be the lowest portion of MEMORY. This part of MEMORY is often reserved for system software; such truncation maps would, to say the least, be inconvenient. To **relocate** the area into which we map, provide an offset quantity, B, which is added to the truncated quantity as illustrated in Fig. 9.2.2 where B = 750.

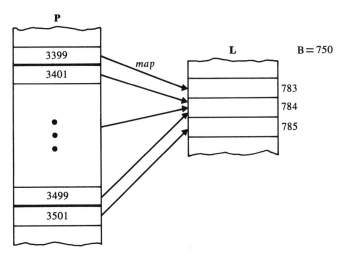

**Fig. 9.2.2.** Suffix truncation and relocation

In general, a truncation and relocation mapping, $map_{TR}$, is given as

$$map_{TR} \; p = B + \lfloor p/d \rfloor \qquad (9.2.3)$$

where B is the base and d is the divisor for truncation. For Fig. 9.2.2 we have

$$map_{TR} \; p = 750 + \lfloor p/100 \rfloor \qquad (9.2.4)$$

and in particular

$$map_{TR} \; 3482 = 750 + \lfloor 34.82 \rfloor = 750 + 34 = 784 \qquad (9.2.5)$$

### Full Truncation Mapping

A practical mapping calculation consists of three activities:

- truncation;
- scaling—multiplication by a constant;
- relocation—addition by a constant.

The truncation divides by the power of a base to provide one cell for a group of keys. Often we would like *several* cells available for each mapping.

Multiplying the result of a truncation by a constant is called **scaling**. It provides several cells for each neighborhood.

In Fig. 9.2.3 we see suffix truncation, scaling by 3, and relocation with the base value of 750.

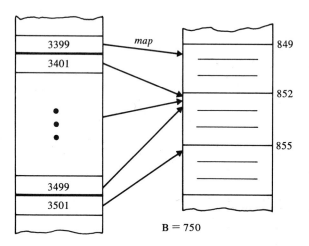

**Fig. 9.2.3.** Truncation, scaling, and relocation

## 9.3 HASHING

**Hashing** is a mapping operation that is not order-preserving—mapped records are no longer in key order.

### Simplest

A logical operation picks out designated bits in the key, extracts them, and compresses them into an address. It may be difficult to see how a useful number can be arrived at thus.

Let us examine keys with a simple prefix and suffix truncation. Both use division: suffix truncation keeps the quotient; prefix truncation keeps the remainder. Prefix truncation drops initial digits from a key. An example is found in Fig. 9.3.1 where a key is a six-digit decimal number. It is truncated by removing the first two and last two digits. This transformation, called $map_{2PS}$, may be written as

$$map_{2PS} = \lfloor p \bmod(10{,}000)/100 \rfloor \qquad (9.3.1)$$

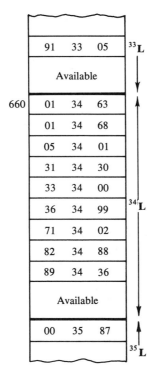

**Fig. 9.3.1.** Hashing by suffix and
prefix truncation

Here dividing by the base squared and neglecting the remainder discards the last two digits. We may subject this result to scaling and relocation as needed. This scheme provides one sublist for each two-digit combination that appears as the *center two digits* of the key. In the figure when the digits 34 are extracted, the resulting sublist is named $_{34}L$.

Records placed in this sublist are ordered in the example. This is not mandatory. Regardless of whether this is done, the entire set **L** will generally end up disordered. This is clear from the figure where the record with key 913305 of $_{33}L$ precedes that with key 013463 in $_{34}L$. Even if a sublist is kept ordered ($_{34}L$ is ordered in the figure), order within the entire list, **L**, cannot generally be maintained.

One might question the efficacy of logical operations because the records end up stored out of order in MEMORY. But notice that records can still be retrieved rapidly compared to other methods so far examined. Logical operations are most useful in setting up sublists that are about equal in size if the extracted digits occur with approximately equal probability—the logical operation is chosen for this quality.

In a random addressing system, uniform sublist size helps avoid misses. For uniformly full (or empty) sublists no sublist is likely to be dense and we are less likely to encounter misses. A sublist which is nearly full is more vulnerable to misses than one which is nearly empty.

The first or last few characters in a key often tend to segregate individuals into groups (such as by states or towns) because they have a meaning with regard to key names. Such groups tend to be nonuniform: California probably has more of our customers than Iowa.

Again, consider given names. There are certainly many more people whose names start with S than with Q. To make uniform classes, we should strip the first few characters from the individual's given name.

### Bit Extraction

Why do we require that the bits we discard are at either end of the key? We can remove or leave arbitrarily assigned bits in any of the digits or characters that comprise the key. For any design, the bits that are extracted are fixed before the actual construction begins. The bits may have no meaning in themselves. The best choice of bits ensures their randomness. That is why this is called **hashing**.

*examples*    To obtain examples of the range of choice, we can do the following:

- extract every fourth bit;
- extract bits 1, 2, 7, 9, 14, 15, 17, 19.

The resulting ensemble of bits is squeezed together and is interpreted as a binary, octal, decimal, etc., number depending on the system we are using.

### Scaling and Relocation

The number obtained from bit extraction can be used directly but would wind up in the lower end of MEMORY. Relocation results in a more suitable starting point. Scaling provides a sublist instead of a single cell for each set of keys.

### Other Simple Methods

Many ways have been thought up to massage a key to make an address that lies within a desired interval in MEMORY.

*multiply*    With large keys, it is possible to break them into two or more subfields. These subfields can be multiplied as binary numbers and the product used as an address. We avoid this method when the multiplier or multiplicand can often be all zeros. Since the point is to obtain a random address, it is unfortunate to end up with address 0.

*add*            Break up the key into subfields, consider them as binary numbers, and add them together discarding the overflow. The result is a random entry. Another alternative is to apply logical functions such as *and* or *or* to these key segments; however, *and* might produce too many zeros and *or* might produce too many ones so that the result would not be as truly random as with simple binary addition of the key subfields.

### Better, More Complicated Methods

There are many other ways to hash a key into a sublist, but these are generally a little more involved.

*division*      If we can arragne to have exactly t sublists, division mod
*mod t*         t is a good way to determine to which sublist the record belongs. If t is prime (not properly divisible by any smaller number), then the method works even better. Division modulus t (or simply *mod* t) takes the key, p and divides by t; throw away the quotient and keep the remainder! Written symbolically, we have for the map, $map_M$

$$p/t = q + r/t; \qquad p = r(\text{mod } t) \tag{9.3.2}$$

or, simply,

$$map_M\, p = r \tag{9.3.3}$$

For example, if we have 47 sublists and are presented with the key 1809, when 1809 is divided by 47, the quotient 38 is ignored and the remainder 23 is kept, thus:

$$map_M\, 1809 = 1809(\text{mod } 47) = 23 \tag{9.3.4}$$

*mid-square*    The mid-square calculation consists of squaring the key and extracting several central digits from the product. This technique is best used when the number of sublists is some power of the base. It is excellent, for instance, for 100 or 1000 sublists.

Starting with our key, 1809, $map_S$ performs the hashing by squaring the number and then extracting two of the center digits, thus:

$$1809^2 = 327\underline{24}81; \qquad map_S\, 1809 = 24 \tag{9.3.5}$$

This hashing consists of squaring the key; dividing by a power of the base, first keeping the quotient; and then dividing by another power of the base, this time keeping the remainder. Call the base B (10 in the example) and we have

$$map_S\, p = (p^2/B^2) \bmod B^2 \tag{9.3.6}$$

*folding and*    Another way to deal with the key is to pick one or more
*addition*    fold points in the key. Digits of the key are added together
   using the boundaries of the fold point for alignment. Any
overflow is discarded. This is hard to see without an example. Take our key
1809 and set the fold point between the eight and the zero. Now the key
breaks into two parts: the first is 18 and the second is 09. Add these together
to obtain 27, which is then the sublist number; thus, for $map_A$,

$$map_A\ 1809 = 18 + 09 = 27 \tag{9.3.7}$$

*algebraic*    There are many other algebraic transformations that can
*transfor-*    be performed upon the key. One of these is to substitute
*mations*    the key into a polynomial. The number produced by the
   polynomial is then reduced by a chosen modulus.

### Review and Contrast

Hashing provides a means of finding the target sublist where
the record is localized by operating on its key. This action is not order-preserving, so that records end up in sublists where they may not occur in key
sequence. The reason for this is so that, given a random group of records,
they will be allocated to sublists uniformly. Truncation does not perform
this action uniformly. Thus if the means for allocating a sublist is the initial
letter of the surname, the Q sublist will be fairly empty, while the S sublist
will be overflowing.

A hashing method can be judged according to how uniform are the contents of each sublist after a number of records are hashed into the list.
Various methods of hashing have been contrasted.† The best methods seem
to be division and the mid-square technique. We are lucky that these are
easy to understand and execute by COMPUTER.

## 9.4 SUBLIST FORMAT; SEARCHING

### Needs of the User

What are the needs of the user with respect to a list? First
we can express them with respect to file alteration—they may be

• examination only;

† V. Lum, P. Yuen, and M. Dodd, "Key-to-Address Transform Techniques," *Communica-
tions of the ACM*, **14**, no. 4, 228–239, April 1971.

- record changes;
- appending and deleting.

Then we can consider them with respect to the order of the file and the order of processing required. Requests might be

- random;
- in key sequence.

To satisfy the user requirements, what alternatives do we have with regard to the organization of the sublist?

- The sublist may be ordered or unordered.
- The sublist may be loose or dense or, as a compromise, semidense.

*matching*     How do we match the sublist organization to the user's needs? It should be clear that some sublist organizations facilitate use. For instance, if we are required to process the file sequentially, then it is uneconomical to keep the file in any other way but in key order. If most of our processing is random, then order is not a necessity. Order is more useful if the records for which we search are frequently absent.

If we are going to append new records, we need cells to hold them. Hence, a dense sublist is useless if much appending is done unless overflow is provided, as discussed in Chapter 10.

Mapping is most useful and efficient when a loose, unordered sublist is maintained and hence when random accessing with appending is the user's need.

### Dense Sublists

Since every cell in the sublist is occupied, appending is not feasible without overflow. To compensate for this, the length of the sublist and the number of records it holds is always known. This structure is ideal when our requests are for retrieval only or for random posting in a file for which alterations are never required.

The structure of the *list* may provide order if the mapping is order-preserving. To be general, we consider all kinds of sublists, even where the mapping is not order-preserving.

*ordered list*     Figure 9.4.1 shows an ordered sublist. It should be clear that the sublist may be kept ordered even though the mapping is not order-preserving.

How do we search the ordered sublist? Examine the key of each record,

$_{33}$L | 3397

$_{34}$L | 3405
3438
3462
3479

**Fig. 9.4.1.** An ordered sublist and    $_{35}$L | 3511
ordered list

continuing either until the desired record is found or until the key encountered is greater than the key sought. Symbolically, write

(a)  $0 \rightarrow i$

(b)  $k \geq L_i$:  *stop*     (9.4.1)

(c)  $i + 1 \rightarrow i$;    *go to* b

*stop* and *go to* have clear meanings. Notice that with (9.4.1) when the record sought has a key greater than that of the last record in the sublist, we use the key of the first record in the *next* sublist to stop the search. Thus in Fig. 9.4.1 when we search for a record with key 3485, we find it greater than 3479 but smaller than 3511.

**ordered**
**sublist**
**only**

If the randomizing technique takes us to a sublist that is kept in order but the entire list is *not* maintained in order, we find a situation such as portrayed in Figs. 9.4.2 and 9.4.3. Notice that the first item of the *next* sublist has a key that is smaller than the last item of *this* sublist. Therefore the next sublist cannot be used to stop searching through this sublist. Figure 9.4.2 is a sublist produced by prefix and suffix truncation, while Fig. 9.4.3 was produced by division modulus 47.

**Fig. 9.4.2.** An ordered sublist of an
unordered list produced by truncation

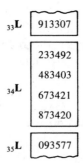

$_{33}$L | 913307

$_{34}$L | 233492
483403
673421
873420

$_{35}$L | 093577

**Fig. 9.4.3.** An ordered sublist of an unordered list produced by division modulus 47

To perform a search of such a sublist we must test to see that we remain in the sublist by checking i against n as in (9.4.2c):

$$
\begin{aligned}
&\text{(a)} \quad 0 \rightarrow i \\
&\text{(b)} \quad k \geq L_{iK}: \quad stop \\
&\text{(c)} \quad i + 1 \rightarrow i; \quad i > n: \quad stop \\
&\text{(d)} \quad go \ to \ b
\end{aligned}
\tag{9.4.2}
$$

**unordered** Figures 9.4.4 and 9.4.5 show unordered sublists, the first
**sublist** developed by prefix and suffix truncation and the second by division modulus 47. Search for a desired record proceeds by checking the key sought for *equal* against the key encountered. This is the only difference with the search of the ordered sublist, so that we have

$$
\begin{aligned}
&\text{(a)} \quad 0 \rightarrow i \\
&\text{(b)} \quad k = L_i: \quad stop \\
&\text{(c)} \quad i + 1 \rightarrow i; \quad i > n: \quad stop \\
&\text{(d)} \quad go \ to \ b
\end{aligned}
\tag{9.4.3}
$$

**Fig. 9.4.4.** Unordered sublists produced by prefix and suffix truncator

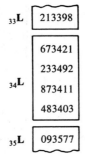

$_{33}$L | 4310

$_{34}$L | 0833
2666
3371
4640

**Fig. 9.4.5.** Unordered sublists pro-    $_{35}$L | 0599
duced by division modulus 47

### Semidense Sublists

For the semidense sublist, the records that have been entered are found at the top of the sublist and empty cells follow these at the bottom. The empty cells are tagged in MEMORY and as before, this is indicated in the figures by @. The semidense sublist provides room for appending new records, and we shall see how this is done while maintaining the sublist structure as semidense in Section 9.5.

*ordered*        For the ordered semidense sublist, the records in the top part of the sublist are kept in order by key as, for example, in Fig. 9.4.6. Presented there is a sublist for which the neighborhood is entered by division modulus 47. To search the sublist, use the algorithm of (9.4.2), except that now we stop for three alternatives:

- We encounter the desired record or one with a higher key.

**Fig. 9.4.6.** Semidense ordered sublist
(modulus 47)

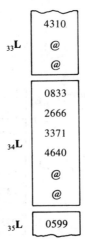

$_{33}$L | 4310
@
@

$_{34}$L | 0833
2666
3371
4640
@
@

$_{35}$L | 0599

- We encounter an empty cell.
- We have examined the entire sublist.

This is indicated symbolically as

$$stop\ for:\quad k \geq L_{iK}\quad or\quad L_{iK} = @\quad or\quad i > n \qquad (9.4.4)$$

**unordered**    If the dense part of the sublist isn't ordered, the sublist of
Fig. 9.4.6 might look like that presented in Fig. 9.4.7. The
search must now proceed accepting only a record with exactly the right key.
We use the algorithm of (9.4.3), the exception being the stopping criterion.
We stop if we find the record, or a space, or have examined the whole sublist,
which is expressed symbolically thus:

$$stop\ for:\quad k = (L_{iK})\quad or\quad (L_{iK}) = @\quad or\quad i > n \qquad (9.4.5)$$

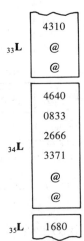

**Fig. 9.4.7.** Semidense unordered sublist (modulus 47)

**Loose**

For the loose sublist, there are fewer records than cells in
each sublist, although some sublists may become full. Unlike the semidense
list, the holes are scattered throughout the sublist. If there are holes at the
bottom of a sublist, they generally originate from cells that were empty and
have not yet been used for records; the holes between records originate from
deleted records. To perform an efficient search, we must distinguish the origin
of each hole. Use the convention that

- A cell from which a record was deleted contains ¢.

- A cell that has remained empty contains @.

Thus we know the bottom of the actual record area of the sublist.

***unordered***    Figure 9.4.8 shows an unordered sublist where the mapping was by division modulus 47. The record in the fourth cell has been deleted.

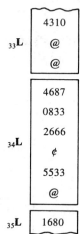

**Fig. 9.4.8** Loose unordered sublist (modulus 47)

Search of such a sublist proceeds by doing a *search equal*, using *esearch*; the search continues if the cell examined contains

- a record with a key different from that sought;
- a deleted record.

$$continue: \quad k \neq (L_{iK}) \quad or \quad (L_{iK}) = \cent \tag{9.4.6}$$

The search stops if

- The record sought is found.
- The cell is empty.
- The entire sublist has been examined.

$$stop: \quad k = (L_{iK}) \quad or \quad (L_{iK}) = @ \quad or \quad i > n \tag{9.4.7}$$

In Fig. 9.4.8, in looking for the record with key 5533, we should not terminate the search because we encounter a deleted record. For an absent record such as that with key 4640, however, we stop at an empty cell rather than examine the whole sublist.

*ordered*    The only advantage of a loose, ordered sublist is that a search for an absent record can be terminated earlier, when we encounter a record with a key lower than that sought. The search rules for a search *equal greater* (*egsearch*) are as follows:

- Continue if the key encountered is smaller than that sought or the cell contains a deleted record.

$$continue: \quad k > (L_{iK}) \quad or \quad (L_{iK}) = \cent \qquad (9.4.8)$$

- Stop if the record is found, one with a larger key is found, an empty cell is encountered, or we have exhausted the sublist.

$$stop: \quad k \leq (L_{iK}) \quad or \quad (L_{iK}) = @ \quad or \quad i > n \qquad (9.4.9)$$

A sample list is found in Fig. 9.4.9.

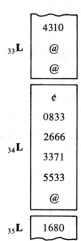

**Fig. 9.4.9** Loose ordered sublist (modulus 47)

**Summary**

*deleting*    It is impossible to delete a record from a dense sublist, for this will create a hole and the list will no longer be dense. When the sublist is semidense, deletion removes a record from the dense part, so that the sublist is no longer semidense. This requires shuffling of the records to maintain a dense and an empty part that are clearly distinguishable. When the sublist is loose, an extra hole does not cause any problem.

***appending***    Again, it is impossible to append a new record to a dense
sublist since there is no room there. Overflow, discussed in
Chapter 10, provides a solution to this. Appending to an ordered list, whether
semidense or loose, may require reshuffling to maintain the desired order.
Appending to an unordered sublist, whether semidense or loose, can be done
using the first empty cell encountered, regardless of its origin, so that no
reshuffling is required.

***contrast***    The characteristics above reveal that no rearrangement is
required for loose, unordered sublists. Hence, if other fea-
tures are acceptable, the loose unordered sublist format is most attractive
when much appending and deleting is done.

## 9.5 DELETING AND APPENDING
### FOR SUBLISTS

### Deleting from Semidense Sublists

Deletion from any semidense sublist must maintain the
characteristic of semidensity. If, additionally, the sublist is ordered, the order
must also be maintained.

***ordered***    Figure 9.5.1 shows an ordered semidense sublist from which
the record with key 2666 is to be deleted. We see on the
right how the remaining records in the dense portion of the sublist are moved
back one cell to replace the deleted record.

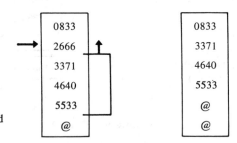

Fig. 9.5.1. Deleting from an ordered
semidense sublist (modulus 47)

To describe this in more general terms, assume the following:

• The record in cell $L_d$ is to be deleted.
• The last cell in the dense portion of the sublist is $L_m$.
• The next cell, $L_{m+1}$, is empty.

$$(L_{m+1}) = @ \quad and \quad (L_m) \neq @ \tag{9.5.1}$$

There will be no need to enter ¢ since records following the deleted record will be moved backward and the cell that used to contain the deleted record will now be occupied by its successor. We indicate the reshuffling of the records symbolically thus:

$$(L_{d+1}) \rightarrow L_d; \quad (L_{d+2}) \rightarrow L_{d+1}; \quad \ldots; \quad (L_m) \rightarrow L_{m-1} \quad (9.5.2)$$

Of course, the last cell that contained a record should now be indicated as empty. There is no need to distinguish between a deleted record and an empty one, so that we have

$$@ \rightarrow L_m \quad (9.5.3)$$

**unordered**     For an unordered semidense sublist, the hole produced by a deleted record must be filled to keep the sublist semidense —otherwise it will become loose; however, the way that we do this need not maintain the present order of the records. In Fig. 9.5.2 we again see the request for deletion of record 2666 (on the left) from an unordered sublist. The records that follow, 3371 and 5533, can be moved back as a group, maintaining the present order as described above, if this is quickest to implement by programming means. Otherwise, the last record may be the only one to be moved backward to the spot previously occupied by the deleted record. Thus, 5533 can be moved back to replace record 2666. In symbolic form we have simply

$$(L_m) \rightarrow L_d; \quad @ \rightarrow L_m \quad (9.5.4)$$

which then replaces (9.5.2) and (9.5.3).

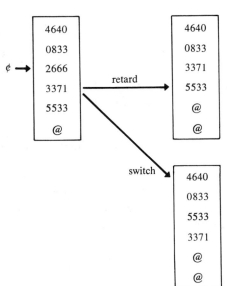

**Fig. 9.5.2.** Deleting from an unordered semidense sublist by retarding or switching (modulus 47)

### Deleting from Loose Sublists

When the sublist is loose, whether it is ordered or not, it is easy to delete a record by tagging the cell as empty with ¢. An example of deletion from a loose, ordered sublist is presented in Fig. 9.5.3. Now ¢ in the second cell of the sublist is no problem during a search. The search procedure not only compares the key in any cell encountered but also checks for ¢ and @. Search continues for ¢ but stops for @.

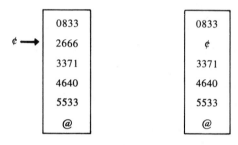

**Fig. 9.5.3.** Deleting from a loose ordered sublist (modulus 47)

Figure 9.5.4 shows an example of deletion from an unordered loose sublist by tagging the empty cell with ¢. Then for both cases, to delete a record from the cell called $L_d$ we have

$$¢ \rightarrow L_d \tag{9.5.5}$$

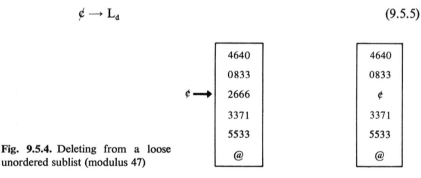

**Fig. 9.5.4.** Deleting from a loose unordered sublist (modulus 47)

### Appending to Semidense Sublists

**unordered**     A new record f′ is added to the first empty cell encountered for any unordered semidense sublist, and that's all there is to it. Since the first m cells of **L** are occupied—the dense part—m + 1 is the number of the first empty cell:

$$f' \rightarrow L_{m+1} \tag{9.5.6}$$

***ordered***       To append a new record to the ordered semidense sublist, reshuffling is required. We found that one of the first applications of graph theory discussed this requirement. Recall how this was done, as presented in Fig. 2.5.8.

### Appending to Loose Sublists

To append a new record to a loose sublist, there are two sites that are available:

- We have originally empty cells marked with @.
- There are also cells which held records which have now been deleted and contain ¢.

We must not overlook the use of the latter, for this would be uneconomical, as should be clear.

***unordered***       For an unordered loose sublist, use the first hole encountered. Thus in Fig. 9.5.5 we wish to add the record with key 8917. Omitted in the figure and the formulas is the search that precedes the insertion. This search must come up with two things:

- the address of the first hole, $L_e$;
- verification that a record with the same key as the new record does not exist in the sublist.

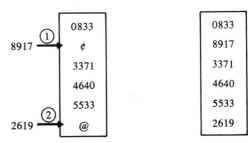

**Fig. 9.5.5.** Appending to a loose unordered sublist (modulus 47)

Then, appending is simply done by placing the new record in the first empty cell. We have then

$$(L_e) = @ \; or \; ¢ \tag{9.5.7}$$

when

$$(L_i) \neq @ \; or \; ¢ \quad \text{(all } i < e)$$

$$f' \rightarrow L_e \tag{9.5.8}$$

Figure 9.5.5 also shows how another record, this one with key 2619, is appended in a cell that has remained empty.

***ordered***    To append a record to the ordered sublist, of course we must maintain this order. When there are several holes in the sublist, it is more efficient to make use of the hole that directly precedes the cell where the record goes. In Fig. 9.5.6 we see a request to append the

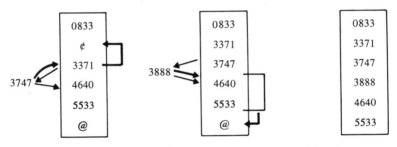

**Fig. 9.5.6.** Appending to a loose ordered sublist (modulus 47)

record with key 3747. The search reveals that the record belongs after 3371 but before 4640. Preceding 3371 is a hole due to a deleted record; following 5533 is a hole that always existed. We could move existing records in either direction and still keep the list ordered. The better choice is to move 3371 into the spot from which the record was deleted and insert 3747 in the hole thus created, as shown in the figure.

When the only existing hole is ahead of the records to be moved, then they must be moved forward. Again, this is shown in Fig. 9.5.6 where record 3888 belongs ahead of record 3474 but behind 4640. Then 4640 and 5533 are both moved forward and 3888 replaces the old 4640, as shown at the right of the figure.

The moving backward criterion and action is indicated symbolically thus:

$$(L_d) = \cancel{c} \quad and \quad f_K < (L_{xK}) \quad and \quad d < x \qquad (9.5.9)$$

$$(L_{d+1}) \rightarrow L_d; \quad \ldots; \quad (L_{x+1}) \rightarrow L_x \qquad (9.5.10)$$

$$f' \rightarrow L_x \qquad (9.5.11)$$

The advance criterion and action is shown thus:

$$L_e = \cancel{c} \ or \ @ \quad and \quad f'_K < (L_{xK}) \quad and \quad e > x \qquad (9.5.12)$$

$$(L_{e-1}) \rightarrow L_e; \quad \ldots; \quad (L_x) \rightarrow L_{x-1}; \quad f' \rightarrow L_x \qquad (9.5.13)$$

### Overlapping Sublists

In the preceding, we have considered sublists as bounded by the entry point from the mapping operation. Why should we restrict ourselves thus? When the entire list becomes full, when the occupancy ratio becomes quite high, some sublists may become full. Instead of recourse to overflow, discussed in Chapter 10, we may consider the use of adjacent sublists. For this to be possible, we require the following:

- Adjacent sublists are contiguous in CORE.
- When order is required, it exists for the entire list, not just the sublist.

Then our search procedure is illustrated in Fig. 9.5.7. Supposing that our entry point is $_iL$. Search proceeds through that sublist into the next, $_{i+1}L$, and then into $_{i+2}L$, and so forth. Search continues until an originally empty cell (tagged by @) is encountered.

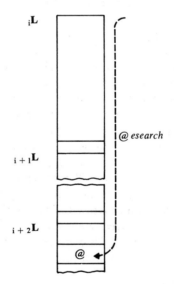

$_iL$

$i + 1^L$

$i + 2^L$

@ esearch

@

**Fig. 9.5.7.** Searching overlapping dublists (modulus 47)

***appending,***
***unordered***
In Fig. 9.5.8 we see two sublists properly segregated. The sublists were mapped using division modulus 47. Records in $_{34}L$ had a remainder of 34 after division; those in $_{35}L$ all had a remainder of 35 after division by 47.

It is desired to append the record with key 1256. First we verify that the record is absent from the list. Since sublists are overlapping, we search

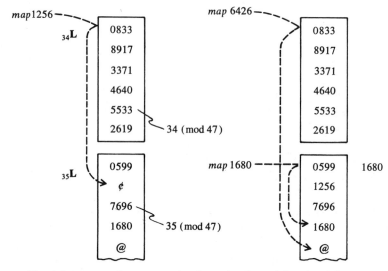

**Fig. 9.5.8.** Appending to unordered overlapping sublists (modulus 47)

through $_{34}$L and $_{35}$L until we hit @ in the fifth cell of $_{35}$L. On the way, we have noted the presence of ¢ in the second cell of $_{35}$L.

Although there is no room in $_{34}$L for the new record, there *is* room in $_{35}$L, and so we place 1256 there. Cells tagged as deleted should always be used before unused cells! True, search lengths will be longer whenever the record sought is mapped into a subsequent sublist since we have to examine subsequent sublists; however, this is the penalty for large occupancy ratios. At the right of Fig. 9.5.8 we see a search using key 6426, which maps to $_{34}$L. Search requires eleven looks: six in $_{34}$L and five in $_{35}$L, before @ is turned up.

Notice that $_{35}$L contains 1256, which is an overflow from $_{34}$L. This does not interfere with any activities related to $_{35}$L. For instance, the figure shows a request for 1680, which is in $_{35}$L. Extra looks at foreign records are required, but eventually the desired one is found.

The user must provide his own criterion for stopping or else he will go on forever. This might be a limit for

- total number of looks;
- number of sublists;
- highest key encountered.

***ordered lists***    The same procedure can be used when the entire list is kept ordered, as illustrated in Fig. 9.5.9. Here suffix truncation is used. Again we find that $_{34}$L is full at the time we wish to append the record with key 3455. Search is continued until we find ¢ in the second cell of

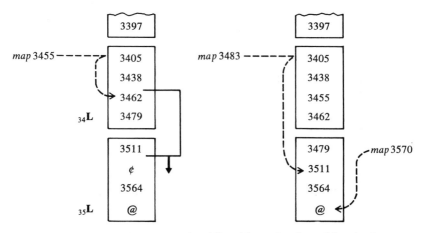

**Fig. 9.5.9.** Appending to an ordered list with overlapping sublists (suffix truncation)

$_{35}L$. This permits us to move records forward across sublist boundaries to make room for the new record. Again, we end up with records in $_{35}L$ that actually mapped into $_{34}L$, and again this does not constitute a problem.

The revised list is shown at the right of Fig. 9.5.9 and the fulfillment of a request for two searches is demonstrated. The search for an absent record, as usual, terminates when a record with a higher key is encountered. For the request for 3483, that record is in $_{35}L$, but at least the search terminates before we reach @. The figure also illustrates a search for 3570.

Of course when overlapping sublists are used, the criterion for stopping the search at the end of a sublist must be removed.

## 9.6 OPEN ADDRESSING

### The Sublist Becomes a List

We have examined various ways in which a sublist structure is imposed upon a list. We have seen that we could divide the list in two ways, so that sublists were not overlapping. In one case, sublists are adjacent; in the other, there is space between sublists. Then we have seen how sublists might overlap so that a search that begins in one sublist might take us into another sublist, or even through that sublist into still another sublist.

Let us now take the overlapping sublist format to its ultimate—completely overlapping—in other words, the whole list becomes our "turf." That is, once we have mapped a key into a cell, the probe that we use might takes us to visit every cell in the list. In fact we can discard the concept of sublists and simply consider

- a mapping that brings us to some cell in the list;
- a probe to follow that is a way or a sequence of visiting each of the cells in the list.

This is **open addressing.**

It is that sequence that we should now examine. In what order do we make our visits?

- We have seen the linear probe that examines each cell according to its position in MEMORY.
- We now look at other alternatives.

### Objections to the Linear Probe

The same technique used to append records is also used for the search. But suppose that the mapping or randomizing technique does not allocate an equal number of keys to each target cell. That is, given a number of keys chosen from the universe of keys at random, more of them will be mapped to some cells than to others.

Suppose, for instance, that more records are mapped to $L_i$ than to other cells in the list:

- The first record to be appended to the list at $L_i$ will actually be placed there if the cell is empty.
- The next record to be mapped to $L_i$ using the linear probe will be placed at $L_{i+1}$.
- The one after will be placed at $L_{i+2}$, and so forth.

Congestion is beginning to take place at $L_i$. Now, a record to be placed at $L_{i+1}$ will find that cell and its successor full, so that it will have to be added further down on the list. Searches for records in this area will necessarily take longer, not only for $L_i$ but also for its neighbors $L_{i+1}$, $L_{i+2}$, etc.

In the literature, techniques have been examined to alter the linear probe, such as a skipping technique. For instance, the probe might skip three cells each time it is invoked. Thus, after appending the first record at $L_i$, the next one would go to $L_{i+4}$. This does alleviate matters since $L_{i+1}$ is no longer in trouble. But as new records are added, destined for $L_i$, we find that $L_{i+4}$ is in trouble, and we have gained little. The best remedy is the random probe, which we now examine.

### Random Probe

The trouble with a linear probe is that full cells seem to cluster together. One way to get away from this is to cause looks during a search to be scattered throughout the entire list area rather than to concentrate in a neighborhood of contiguous cells. We still have a neighborhood but it is dispersed through MEMORY.

***method***     When a look results in a miss, the linear probe takes the next physical location for its next look. The random probe generates a new location that is, to some degree, randomized. We do this with a **pseudorandom number generator**. This is prefixed *pseudo* because although the numbers produced are random, they are repeatable; each time we use the generator from a given starting point, it takes us to the same cell next. The pseudorandom number generator operates upon a location in a manner that we symbolically designate with the operator *probe*. Thus if we start at *map* k, we apply the random generator *probe* and come up with a new address *probe map* k.

Then in conducting a search we go through this sequence of the looks:

$$map \text{ k}, \quad probe \text{ } map \text{ k}, \quad probe^2 \text{ } map \text{ k}, \quad probe^3 \text{ } map \text{ k}, \quad etc. \qquad (9.6.1)$$

where the superscript describes the number of times the probe was made.

***example***     Consider a method of moving around through the list so that we encounter cells no more than once and all cells exactly once before returning to the original cell. One operation that does this multiplies the list address in binary form by $\frac{5}{4}$, casting off the overflow bit when it is 1. This is simple to program without multiplications. We shift the binary number two places to the right to produce one-quarter of the address. This is added to the original address to produce $\frac{5}{4}$ times the original number. Overflows during the addition are ignored.

***order***     It is virtually impossible to keep a list in order when the random probe is used. Hence we can assume that a search is confronted with an unordered list. The rules for search are

(a)   $j = 0$

(b)   $(probe^j \text{ } map \text{ k})_K = \text{k}$   *or*   $(probe^j \text{ } map \text{ k})_K = @:$   *stop*     (9.6.2)

(c)   $j + 1 \rightarrow j;$     *go to* b

Figure 9.6.1 shows a search graphically. Solid lines indicate the sequence of the cells in MEMORY; dashed lines show the sequence of our looks. The

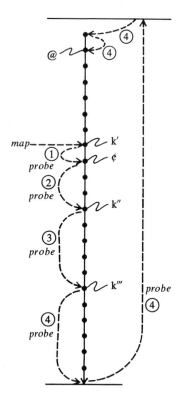

**Fig. 9.6.1.** Open addressing

first look at *map* p finds p'. The second at *probe* finds ¢. The third look takes us to *probe²* *map* p occupied by p''. The fourth look takes us back near the beginning of the list to find @, an empty cell at *probe³* *map* p; we terminate our search here.

### Quadratic Probe

The quadratic probe is another effective pseudorandom probe. Given a starting point in the list, *map* k, a quadratic function of the look number j is added to this. Then j is the number of times we use the pseudorandom operator. Said another way, j is the exponent to which we raise *probe*$_Q$.

Originally the generator worked as a quadratic function. That is, we added bj and j² thus:

$$probe_Q^j \; map \; k = map \; k + bj + j^2 \qquad (j = 0, 1, \ldots) \qquad (9.6.3)$$

Then Maurer† found that a simplification would work just as well. The simplification consisted of omitting bj—simply add $j^2$ to the cell address where j is the ordinal number of the probe. Thus:

$$probe_Q^j\ map\ k = map\ k + j^2 \tag{9.6.4}$$

To use this system, the simplified quadratic probe, examine the list in this fashion:

- Map to a particular cell $L_i$.
- Look at the next cell, $L_{i+1}$.
- The next look is four away from the original cell, at $L_{i+4}$.
- The next is nine away from the original cell, at $L_{i+9}$, etc., so that we have

$$\begin{aligned} probe_Q\ map\ k &= map\ k + 1; \\ probe_Q^2\ map\ k &= map\ k + 4; \\ probe_Q^3\ map\ k &= map\ k + 9 \end{aligned} \tag{9.6.5}$$

The method of (9.6.4) takes us through half of our list without any difficulty. This leaves a number of cells unexamined. There are improvements in the algorithm described in the literature that permit us to visit *every* cell in the list. When occupancy ratios are so high as to necessitate such a thorough examination of the list, however, the entire technique becomes inefficient and should be avoided.

**Efficiency**

One criterion for comparing organizations is the number of looks.

The efficiency of any of the methods proposed can be determined by the number of looks it takes during search to find the desired record or to verify that it is absent. It can be made independent of the size of the list and solely dependent on the occupancy ratio, OR.

$$OR = num\ \mathbf{f}/num\ \mathbf{L} \tag{9.6.6}$$

where **f** is the file contained in the list **L**.

A few assumptions are required:

- The universe is a linear distribution—placing a record in any cell is equally likely.

---

† W. D. Maurer, "An Improved Hash Code for Scatter Storage," *CACM*, **11**, no. 1, 35–38, January 1968.

- The sublist is open—some keys map into every cell in the list.
- The key universe is statistically large.
- We find the *average* number of looks for random choices of keys.

The average number of looks, $\hat{L}$, for various values of the occupancy ratio for the linear probe is presented in tabular form in Table 9.6.1, as discussed by Morris.† It also contains a similar table for the random probe, from Morris as well.

**Table 9.6.1.** Contrast of Linear and Pseudorandom Probe for Various Occupancy Ratios

| | Looks | |
| Occupancy Ratio | Linear Probe | Pseudorandom Probe |
| --- | --- | --- |
| 0.0 | 1.00 | 1.00 |
| 0.1 | 1.06 | 1.06 |
| 0.5 | 1.50 | 1.39 |
| 0.75 | 2.50 | 1.83 |
| 0.90 | 5.50 | 2.56 |

Why does the random probe provide such an improvement? For multiple appending, after a few misses are resolved by the linear probe, new records tend to clump together in consecutive cells of the list. After a miss at $L_i$ during appending, there is a higher probability of a miss at $L_{i+1}$. Clumping breeds clumping! This does not happen in the random probe.

**Note**

A probe generally continues to completion:

- For appending, an empty cell is found.
- For search, the record is found or proved absent.

Exceptions may arise of which we must be wary:

- There are no empty cells; OR $= 1$.
- Although the sought record is absent, the search may tend to continue, such as for search *greater* in a list where all records are *less*.

† R. Morris, "Scatter Storage Techniques," *CACM*, **11**, no. 1, 18–44, January 1968.

The programmer must make sure to terminate by some criterion such as

- a cell count;
- an address check.

## PROBLEMS

9.1 For mapping, distinguish among
   (a) the key population, **p**;
   (b) the file, **f**;
   (c) the list, **L**.

9.2 Distinguish the four quantitative alternatives for mapping:
   (a) one-to-one,
   (b) many-to-one,
   (c) one-to-many,
   (d) many-to-many.
   Why do we need a many-to-one mapping?

9.3 Neighborhoods have alternatives as to overlap. What are they, and which might we choose from a list and why?

9.4 For truncation:
   (a) What is it?
   (b) When would we employ
      - prefix truncation?
      - suffix truncation?
      - both?
   (c) Express *all* the above mathematically.

9.5 What is a *hit* and a *miss*?
   (a) How are these defined when the record is present?
   (b) How are these defined when the record is absent?
   (c) How are these defined when looking for room?
   (d) When do the needs above arise?

9.6 Consider an application with these parameters:

$$num\ \mathbf{p} = 100\mathrm{K}; \quad num\ \mathbf{f} = 800; \quad num\ \mathbf{L} = 1\mathrm{K};$$
$$len\ f_{i\mathrm{K}} = 7; \quad len\ f_i = 120; \quad [\mathbf{L}] = 23{,}000$$

Derive a mapping using the technique of
   (a) prefix truncation;
   (b) suffix truncation;

(c) both, equally.

(d) For each, where do records with these keys go?

$$1, 234, 567; \quad 2, 323, 232; \quad 4, 443, 322; \quad 67867867;$$
$$00000017; \quad 1, 230, 000; \quad 0, 234, 560; \quad 9, 498, 863;$$
$$9, 000, 006; \quad 1, 093, 901$$

9.7 Explain hashing by bit extraction.

9.8 The file, **f**, may be seen as a sample from the population, **p**. Why is it so important that this be a *random* sample? If it is not random, how does this affect the mapping we use? What is the result of a "poor" sample and a "good" mapping?

9.9 (a) Give a verbal description of mapping by division modulus $m$.
(b) How does one choose $m$? Why?
(c) What is the relation between $m$ and *num* **L**?

9.10 Correlate these two sets of alternatives:
(a) file alteration;
(b) sublist organization—order and density.

9.11 Explain how an ordered sublist does *not* imply an ordered list.

9.12 When and why does a loose sublist require that we distinguish between an empty cell and a cell with a deleted record?

9.13 (a) When a list is subject to revision, what sublist format is preferable?
(b) Why?
(c) Explain appending and deleting in such a case.

9.14 If *order* is also a requirement, what are the answers to Problem 9.13?

9.15 (a) Explain the principle of overlapping sublists (OS).
(b) May an OS be loose? dense? ordered? unordered?

9.16 For an OS, *this* sublist may contain records from the preceding subfile; records from *this* subfile may be in a subsequent sublist.
(a) Explain how a search begins for these cases for the ordered and the unordered lists.
(b) Explain how the search ends for ordered and unordered lists.

9.17 For open addressing (OA):
(a) What is it?
(b) How is a neighborhood defined?
(c) What is a *hit* and a *miss* for OA?

9.18 For OA probes:
(a) What is a linear probe, and what are the objections to it?
(b) What is a pseudorandom probe (PRP)?
(c) Why is the PRP *not* really random?

9.19 (a) Explain the quadratic probe (QP).
  (b) Write a QP in a familiar program language.

In the following problems, consider an eighty-byte record with an eight-byte character numeric key. Map with *map*, which is division modulus 101.

9.20 N consists of nonoverlapping ordered semidense sublists, in general, called $N_i$. We have

$$[N] = 50,000; \quad num\, N_i = 5; \quad num\, N = 505$$

Show the pertinent position of N after the records of Table 9.P.1 have been entered there.

**Table 9.P.1**

| |
|---|
| 9945 |
| 16004 |
| 24792 |
| 63880 |
| 56103 |
| 81960 |
| 91048 |
| 65394 |
| 84383 |
| 2370 |
| 99230 |

9.21 For N, do these actions in this order and show the result:
  (a) Delete 24792.    (d) Delete 9945.
  (b) Delete 63880.    (e) Add 96199.
  (c) Add 20147.       (f) Add 1764.

9.22 M consists of nonoverlapping unordered sublists $M_i$ for which

$$[M] = 40,000; \quad num\, M_i = 5; \quad num\, M = 505$$

Repeat Problem 9.20 for M.

9.23 Repeat Problem 9.21 for M.

9.24 O consists of overlapping unordered sublists $O_i$ for which

$$[O] = 66,000; \quad num\, O_i = 4; \quad num\, O = 404$$

Repeat Problem 9.20 for O.

9.25 Repeat Problem 9.21 for O.

9.26 For **M** of Problem 9.21, show how to search for
    (a) 20147,                  (e) 33679,
    (b) 2370,                  (f) 84383,
    (c) 99230,                (g) 66101,
    (d) 63880,

9.27 For **M** of Problem 9.23, repeat Problem 9.26.

9.28 For **O** of Problem 9.25, repeat Problem 9.26.

# Overflow

## 10.1 INTRODUCTION

### What Is Overflow?

**Overflow** provides a secondary list of cells to supplement the main list when the latter becomes full. We designate overflow with the letter **H**. The secondary list is needed when any appending is being done for a dense main list or for excessive appending to a loose main list. A combination of appending and deleting creates holes that may or may not be reusable. When there is no room in the main list, we reach a stalemate when trying to append a new record if some other area were not provided.

Overflow, **H**, may be categorized in two ways. One of these is the position of the overflow area with regard to the main area:

· It may be adjacent.
· It may be distant—in another part of MEMORY or on a different medium.

The other way to classify the overflow area is according to how it is used with respect to the sublist structure:

· If there is just one overflow area to be used regardless of where in the main list the record belongs, it is called **global overflow**.
· When one overflow area is provided for each sublist at the lowest level (e.g., subsublist in a two-level directory system), this is called **local overflow**.

279

- One area of **intermediate overflow** is provided for each fraction of the list, but for something greater than the lowest level of sublist (e.g., intermediate overflow is provided on the sublist level for a two-level directory system).

### Local Overflow

There is one local overflow area for each low level sublist. Designate the sublist as $_iL$ and the corresponding overflow area as $_iH$. One way to do this is shown in Fig. 10.1.1. The overflow area H is divided into n

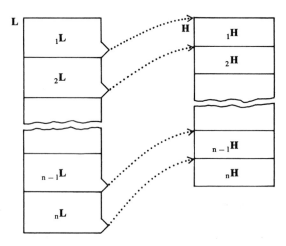

**Fig. 10.1.1.** Local overflow provides an overflow area, $_iH$ for each sublist $_iL$.

individual areas, $_1H$ through $_nH$. The list is similarly divided into n sublists. When a sublist is full, further records are entered into the corresponding overflow area. When the overflow area is distant, some means for locating it is required. The figure shows a pointer attached to each sublist that locates the overflow area for that sublist.

It is possible to locate the overflow area implicitly. Thus if the location of the overflow area and the allocation for each sublist is known, no pointer is required. For example, suppose the overflow area begins at H where 100 bytes are allocated for each local overflow area. Then local overflow for $_1L$ is also at H; the overflow for $_2L$ is at H + 100; similarly, overflow for $_4L$ is at H + 300, etc.

*semidense*   We are already familiar with the semidense list and sublist, as shown in Fig. 10.1.2. Here the sublist, $_1L$, is considered to consist of two parts:

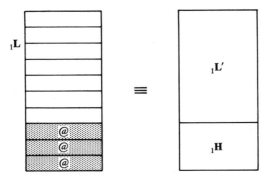

**Fig. 10.1.2.** The semidense sublist provides floating local overflow.

- The dense part, $_1L'$, is at the top of the sublist.
- The empty part, $_1H$, is at the bottom and provides room for overflow.

The boundary between the two floats. Whereas the sublists and local overflow for them presented in Fig. 10.1.1 had fixed boundaries, for Fig. 10.1.2 the demarcation between the two areas depends on the occupancy ratio.

### Global Overflow

One global overflow area serves all neighborhoods or sublists, as shown in Fig. 10.1.3. At the left of the figure we see the main list, L, kept by any of the disciplines we have examined previously. When there are no longer cells available in a given sublist such as $_1L$, a single pointer for that sublist is employed to point to a cell in overflow in which a new record is

**Fig. 10.1.3.** A global overflow, H can serve all sublists.

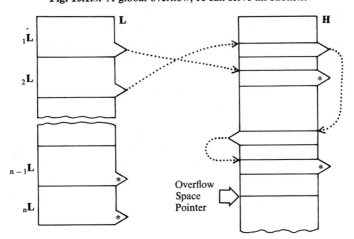

appended. Each overflow cell has a pointer field by which it may be chained to other cells in overflow. Thus besides its own cells, $_1L$ employs three additional cells in **H**. The last of these has a terminal pointer, * in the pointer field as a flag to indicate search termination.

$_2L$ has required only a single overflow cell, and a sublist pointer links to that cell in overflow. $_nL$ does not require recourse to overflow; hence the sublist pointer contains a terminal symbol.

*subfiles*        For explanation, consider a sublist using overflow to consist of two parts:

- the **physical sublist**, which is determined by those cells which occupy the physical sublist boundaries;
- the **logical sublist**, or better, the **subfile**, which includes those records in the sublist proper and those records belonging to the subfile but which are contained in overflow.

For example, in Fig. 10.1.3:

- The sublist, $_1L$, consists of the cells that comprise $_1L$ and the records contained therein.
- The first subfile consists of records in $_1L$ and the three records in overflow.

*monitoring*      To append a new record, first the sublist is searched, and
*global*          (if full) then the overflow cells belonging to the subfile are
*overflow*        examined. In other words, we search the subfile exhaustive-
*space*           ly. We can only append a record to the subfile if it is absent.

Having verified this, we now need space for the record. Overflow space might be monitored by an overflow space pointer, as seen in Fig. 10.1.3.

To append a new record, follow the overflow space pointer to free space in global overflow. Withdraw sufficient space for the new record and adjust the overflow space pointer accordingly. A terminal pointer is entered into the new cell and the old last cell of the subfile is pointed to this new cell.

Similarly, to add a new record to the file in $_nL$, use overflow only when cells were no longer available in that sublist. Then the sublist pointer would be pointed to the new cell just obtained.

*order*           Order for records in overflow exists or does not exist accord-
                  ing to the structure of the sublist that overflow serves. For an ordered sublist, order is maintained throughout the subfile. Thus the records in the physical sublist are those with lower keys; records with higher keys are chained together in overflow and order maintained within the linked list.

The subfile consists of two parts: the sublist proper and the cells in overflow associated with the subfile. For an unordered subfile, when a new record is added, it is immaterial where that record goes. When there is room in the sublist, the record is placed there. When the sublist is fully occupied, a new cell is obtained in overflow and the record is placed there. To make sure that the record remains associated with the subfile it is necessary to place it in the overflow chain for that subfile. Again, it is immaterial what position the record occupies. Then the simplest thing to do is to find the overflow chain belonging to the subfile. The cell containing the new record may be placed at the beginning or the end of that chain (or anywhere else in the chain), whichever is more convenient.

When a new record is appended to an ordered subfile, order must be maintained. Assume here that the subfile is larger than the main sublist and thus occupies part of overflow as a linked list there. A search reveals whether the new record belongs in the sublist proper or in overflow. We examine these two cases in that order.

For the first case shown in Fig. 10.1.4, the record at **NEW** belongs between two records with keys 3723 and 3738 contained in the sublist. This case is similar to the manipulation required for the semidense sublist. The result of doing the actions below is illustrated in Fig. 10.1.5:

**Fig. 10.1.4.** Using double suffix truncation, sublist size of 5, and global overflow, add a new record to go  in the sublist proper.

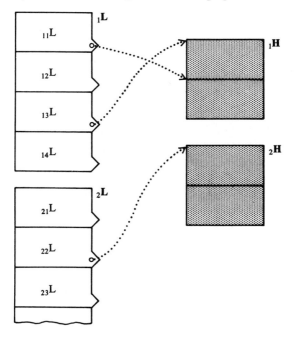

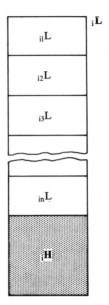

**Fig. 10.1.5.** Showing the result of adding the record as in Fig. 10.1.4; now add another record to go into the overflow chain.

1. Find the cell where the new record belongs in the sublist.
2. Temporarily store the record now in the *last* cell of the sublist.
3. Move records to their successor cells in the sublist until the record in the needed cell has been duplicated in the cell below it.
4. Enter the new record in the cell freed by the action above.

This leaves an evicted record from the end of the sublist to be taken care of. An overflow cell is commandeered using the space pointer. The evicted record is placed there. This new cell is pointed at the beginning of the overflow chain for the subfile since that chain is still in order and this record should precede it. Finally, the last cell in the sublist is pointed at the new cell of overflow containing the evicted record. This action can be done at step 1 above to obviate saving in a temporary cell.

The second case is also shown in Fig. 10.1.5. The new record 3790 does not belong in the sublist proper but should be someplace in the overflow chain for the subfile. As soon as this is determined, it is possible to obtain a new overflow cell for the record and place it there. Search continues through the overflow chain until the record that should precede this one is found. We shall probably have to find the record to follow this one first and then determine its predecessor. Insertion proceeds in typical linked list fashion:

1. Change the pointer in the predecessor to the new cell containing the new record.
2. Point the new record at its proper successor.

### Intermediate Overflow

Intermediate overflow is applicable only for lists comprised of sublists of at least two levels. For instance, in the two-level directory system we divide our original list into sublists; then each sublist is divided into subsublists. For the multilevel directory file, each subsublist is further divided into subsubsublists, and so forth.

As the adjective implies, **intermediate overflow** provides overflow somewhere *between* the local and the global level. Consider a two-level system:

- Global overflow is used by all sublists and subsublists.
- Each local overflow is used by only a single subsublist.
- Each intermediate overflow is used by a single sublist—by all the subsublists that comprise that sublist.

*contiguity*  Overflow areas are also classified according to whether they are adjacent to the sublist that they serve or distant from it.

*need*  For DIRECT ACCESS STORAGE DEVICES, where it takes considerable time to position heads to a new cylinder, intermediate overflow provides a solution necessitated by the physical constraints of the medium.

## 10.2 DIRECTORY FILES WITH OVERFLOW

### Appending and Deleting

As described in Chapter 7, the directory file uses dense sublists and a dense directory. As such, it is impossible to perform appending and deleting: there is no room for the former and the latter would create holes that are intolerable. The way to cope with this deficiency is to provide overflow. Any of the three kinds of overflow discussed in Section 10.1 could be used to make appending and deleting possible. A further safeguard is to provide a combination of two kinds of overflow.

The directory file, as discussed previously, required ordered, dense sublists. We can handle loose sublists; do they still need to be ordered? An argument for order arises when the file is accessed sequentially for posting or for periodic printout. Either of these provide a strong objection to loose organization and this must be carefully considered in the file design. Otherwise, having found the sublist, there is no reason to keep that sublist ordered if we are simply trying to locate a single record and improve search time—an ordered sublist, as we know, does *not* provide better search time.

### Semidense Sublists

Contiguous local overflow amounts to a semidense sublist. Sublists kept in this format allow for appending and deleting with concommitant reorganization to maintain the semidensity.

Figure 10.2.1 shows two directory entries and the sublists to which they pertain, where semidense sublists are used. Again we distinguish between the physical sublist and the logical sublist or subfile.

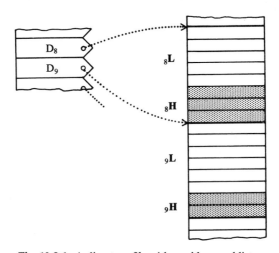

**Fig. 10.2.1.** A directory file with semidense sublist

**subfile**    The extent of the subfile is determined by the directory entry and namely by its key field. It should be clear that records in the file have keys greater than the key field of the previous directory entry and less then this one. Corresponding to the ith directory entry key, $D_{iK}$, is the sublist $_iL$; a general record in this sublist $_iL_j$ has the key $_iL_{jK}$ and we have

$$D_{iK} < {_iL_{jK}} \leq D_{i+1K} \qquad \text{(for } j = 1 \text{ to n)} \qquad (10.2.1)$$

Actually the boundary between the subfile and its local overflow is reached during the search when a cell containing @ is encountered, as illustrated in the figure.

**sublist**    Again, as illustrated in Fig. 10.2.1, the sublist boundary is
**boundary**    usually conveyed by the directory entry:

• The beginning of the sublist is found in *this* entry.

- The end of the sublist for the contiguous sublist structure is determined by the beginning of the *next* sublist.
- Another way to provide this is implicitly, where the number of cells in the sublist is known.

Ways to search, append to, or delete from the semidense sublist have been examined in Section 9.4, and these are applicable here.

### Local Overflow

For local overflow (other than the semidense type), a distant overflow area, $_iH$, is provided for each sublist, $_iL$. The use of the sublist for search, appending, and deleting should be clear. The question now arises as to how to find $_iH$ during these activities.

*sublist*     One way to find the local overflow area is to provide an
*pointer*     overflow pointer in the last cell of each sublist. This additional field in that cell need not appear for other cells in the sublist, as illustrated in Fig. 10.2.2. During use, the sublist is examined until

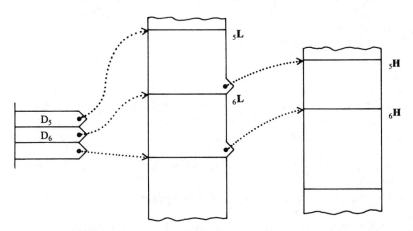

**Fig. 10.2.2.** A directory file with implicit local overflow

it is exhausted. This is determined implicitly, knowing the number of cells in the sublist, or explicitly, from the next directory entry. The additional field of this last cell is then examined:

- A terminal pointer or a flag shows that local overflow has not been used.
- Otherwise this field *is* a pointer to local overflow.

***directory*** Another technique illustrated in Fig. 10.2.3 expands the
***pointer*** directory entry, $D_i$, placing there an additional pointer,
$D_{iH}$, which *is* the overflow pointer. After having exhausted
the sublist, search continues using this pointer. It may be flagged to convey
if there are actual overflow records.

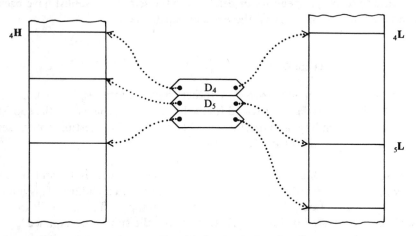

**Fig. 10.2.3.** A directory file with local overflow by directory pointer

### Global Overflow

At the other extreme we find global overflow, where every
sublist is served by a single global overflow area, **H**, as illustrated in Fig.
10.2.4. The subfile consists of the records contained in the sublist and those
found in global overflow. The two means described above for finding the
first cell in overflow belonging to a given subfile can also be employed here:

- An extra field in the last cell of the sublist can point to the first overflow
  record.
- The directory can contain a field for this purpose, as in Fig. 10.2.3.

Note that *no* cell in global overflow is assigned to a specific subfile.

***space*** Whereas, in local overflow, the overflow sublist is small and
***pointer*** can be exhaustively searched quickly so that it is possible
to use an empty cell symbol to designate available space,
this does not prevail for global overflow. To keep track of available space in
global overflow, an overflow space pointer is generally required.

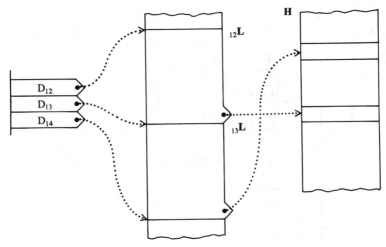

**Fig. 10.2.4.** A directory file with global overflow

### Intermediate Overflow

If we have at least two levels of directories, intermediate overflow is an ideal solution, especially for files kept on DASD volumes. Then a subdirectory, a sublist, and an intermediate overflow area can be kept in the same proximity, as illustrated in Fig. 10.2.5. Then the subsublists that comprise the sublist all make use of this intermediate overflow area.

How do we find the first overflow record?

- The last cell in the subsublist may point to it, as shown in Fig. 10.2.5.
- A directory pointer of the type presented in Fig. 10.2.3 may be used.
- A separate directory entry field may be used for the overflow portion of the subsubfile.

*separate directory entry for overflow subfile*  This last technique is important because it is used in IBM's Index Sequential Access Method or, simply, ISAM. The technique is illustrated in Fig. 10.2.6.

For each subsubfile, the directory contains two entries:

- The first entry
  * points to the starting location of the sublist;
  * contains the key of the last record in the sublist proper.

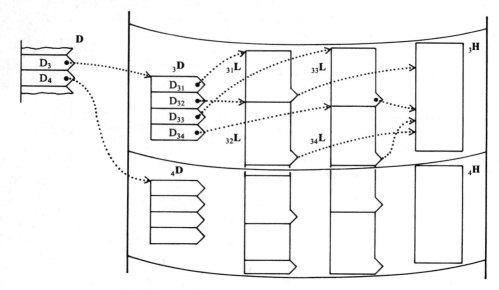

**Fig. 10.2.5.** A directory file with intermediate overflow with local pointer

- The second entry
  - \* contains the location in intermediate overflow where the *first* record of the overflow subsubfile is found;
  - \* contains the key of the last record in the overflow linked list subsubfile.

Thus to use the directory for search, entries are examined in order. Since there are two entries for each subsubfile, the entry reached determines whether we examine the subsublist or the remainder of the subsubfile in overflow. Since each subsublist is the same size, it is possible to know when search is complete because of the number of records encountered. When searching an overflow subsubfile, it is the pointer that determines either the next record or that no further records are provided.

*supplementary* When a file requires frequent appending, it is truly cata-
*global* strophic when there is no further room in the subsublist *or*
*overflow* in intermediate overflow. The program that is running must
be terminated in this case. To provide for temporary recovery, global overflow can be used *in addition to* intermediate overflow, again as illustrated in Fig. 10.2.6. Here, intermediate overflow, $_iH$, provided for the subfile $_iL$ has become exhausted before some of the new records for that subfile had been appended. We now have recourse to the global overflow area, $H$, at the bottom of the figure. The heavy line separating it from the

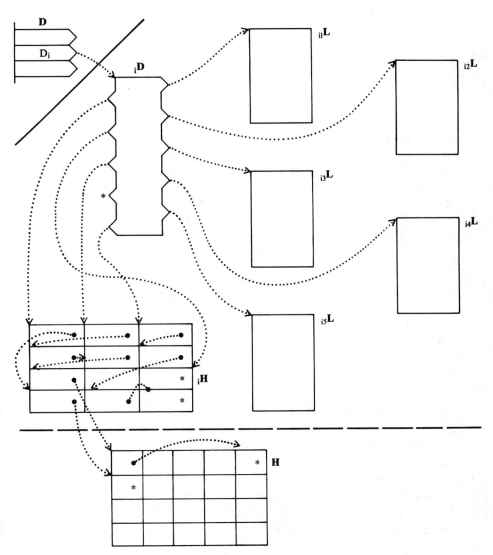

**Fig. 10.2.6.** A directory file with intermediate overflow with directory pointer

upper part of the figure conveys that it is located at a distance from the sublist proper. As a consequence, using this global area during normal processing is uneconomical because of the delays in going back and forth to reach it. When this condition arises, the directory file should be reorganized.

## 10.3 MAPPING WITH OVERFLOW

### Need

We have seen that appending new records to a list where mapping is used causes many misses during searches for high occupancy ratios. When the entire list is used—open addressing—clumping and clustering results because several records are mapped to the same cell. We have seen that this effect is reduced by providing a pseudorandom probe instead of a linear probe. Still, records destined for a particular cell are now placed in other cells of the sublist, which interferes with finding or adding records which are mapped to this new cell. This interference can be eliminated if records that map to an occupied cell are not placed elsewhere in the list but are put rather in an overflow area. Then they will not occupy other cells into which other records map, and they will not interfere with mapping to other cells.

We now require an extra area for records that are not held in the list; this buys us a lower occupancy ratio within the main list. That improves the number of looks for searching and appending but requires extra storage. Undoubtedly the extra storage is worth the improvement, as we shall see.

### Global Overflow, Small Sublist

As before, for each target mapping, we provide a sublist, and this sublist contains n cells, where n is greater than one. This sublist may be semidense or loose, ordered or unordered. As shown in Fig. 10.3.1, the last cell in the sublist has an additional field, a pointer. When the list is initialized, the pointers are all set to *, and @ is entered into each empty cell.

As long as the sublist has space, new items that map there are placed in the sublist. When we run out of space, a new cell is obtained from the global overflow area, **H**, using the space pointer technique. The pointer in the last cell of the sublist is set to point to the new cell. The new cell in turn contains a pointer field, which is set to *.

As additional new records are mapped into this sublist, they are placed in overflow, wherein they form a linked list as exemplified by $_{34}L$ in Fig. 10.3.1.

### Global Overflow for the Unit Sublist

Since we provide overflow, why not try to keep most of our records in overflow? That is, provide a skeletal list, as shown in Fig. 10.3.2. Each sublist consists of a single cell. Now, if a sublist is unused—no record

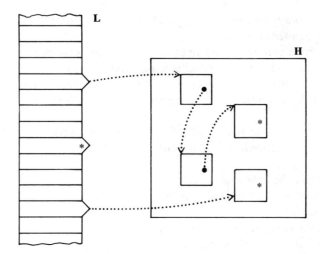

**Fig. 10.3.1.** Mapping with global overflow, local pointer

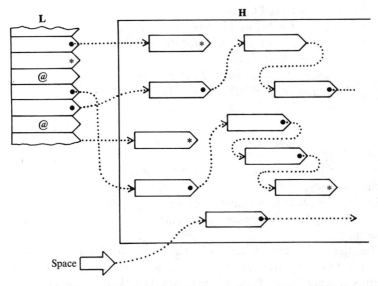

**Fig. 10.3.2.** Mapping with global overflow for unit size sublist

maps to it—than at most one cell goes to waste. Further, all cells in the list are of the same size—each has an overflow pointer that may or may not be flagged as terminal.

For any given sublist, one of these conditions then exists:

- The subfile is empty and hence the cell (sublist) contains @ and a terminal pointer, *.

- There is only one record in the subfile, and it is found at the sublist cell whose pointer contains *.
- The subfile contains more than one record, the first of which is found in the cell and the next is contained in overflow, in a cell pointed to by the sublist pointer—other records of the subfile are found in overflow cells in the linked list stemming from the original sublist cell.

*search*    To search a unit sublist with global overflow, the following procedure is followed:

A. Examine the sole sublist cell:
  * For a *hit*, we find an empty cell (append) or the desired record (search).
  * For a *miss*, we find some other record.
B. Examine the overflow pointer
  * For a *hit* (absent record or append), it is *.
  * For a *miss*, it points to the first cell in the linked list.
C. Examine the linked list cell
  * For a *hit*, it is the desired record.
  * For a *miss*, it is some other record or a delete indicator.
D. Examine the link pointer
  * For a *hit* (absent or append), it is * and we stop.
  * For a nonterminal link, *go to* 3.

### Appending

The technique for appending is similar for both unit and multicell sublists.

First, examine the main sublist to see if there is space. If so, enter the record there. This may require reordering or compressing, according to the format of the sublist.

When the main sublist is full and equal in size to the subfile, its overflow pointer is terminal. Get space from overflow using the overflow pointer, updating it when done. If no order is maintained, the new record may be dropped into this overflow cell; otherwise some reorganization is in order. The sublist is pointed to the new cell, into which a terminal pointer has been placed.

When a portion of the subfile is already in overflow, a new cell is obtained, the record is put there, and it is inserted into the chain. When order is not maintained, the new cell with the record goes to the end. This cell has a terminal pointer in it to stop a search. To keep order, pointers in the linked list are reset and the sublist may *also* require reordering.

### Delete

To delete an item from the sublist proper, use methods described in Chapter 9, by either entering ¢ or reorganizing the list to provide @ at the bottom.

When there is only one record in the linked list area, this record can be deleted by simply setting a terminal pointer into the sublist link pointer and releasing the space.

When there are several records in the linked list area, one of them is deleted by linking around it and releasing the cell that the record occupied.

### Contrast

To contrast the technique described herein with other methods, we examine not only the number of looks required to perform a search but also the additional storage required for overflow. Morris[†] and Johnson[‡] have discussed the number of looks required. They have also shown how the additional storage can be determined. The various quantities concerned are presented in Table 10.3.1. We require new definitions. For the simple occupancy ratio, substitute list occupancy ratio, LO. We have these definitions:

- **List occupancy**, LO, is the ratio of file size to list size;

$$LO = num \; \mathbf{f}/num \; \mathbf{L} \tag{10.3.1}$$

- **Storage** is the union of the list and overflow area;

$$S = \mathbf{L} \cup \mathbf{H} \tag{10.3.2}$$

- **Storage size**, SS, is the number of cells in storage;

$$SS = num \; \mathbf{S} = num \; \mathbf{L} + num \; \mathbf{H} \tag{10.3.3}$$

- **Storage occupancy**, SO, is hence the ratio of the file size to storage size;

$$SO = num \; \mathbf{f}/num \; \mathbf{S} \tag{10.3.4}$$

- **Overflow ratio** is the ratio of overflow to list cells;

$$num \; \mathbf{H}/num \; \mathbf{L}$$

† R. Morris, "Scatter Storage Techniques," *CACM*, **11**, no. 1, 38–44, January 1968.
‡ L. R. Johnson, "An Indirect Chaining Method for Addressing on Secondary Keys," *CACM*, **4**, no. 5, 218–222, May 1961.

Table 10.3.1. Storage Required for the Overflow Probe, and a Contrast of Search Looks for Other Probes

| List Occupancy | Overflow Ratio | Storage Occupancy | For a List of 1000 Cells | | | Overflow Probe | Search Looks | |
| | | | File Size | Overflow Size | Storage Size | | Random Probe | Linear Probe |
|---|---|---|---|---|---|---|---|---|
| 0.1 | 0.0048 | 0.0995 | 100 | 5 | 1005 | 1.05 | 1.06 | 1.06 |
| 0.5 | 0.1068 | 0.4518 | 500 | 107 | 1107 | 1.25 | 1.36 | 1.45 |
| 0.75 | 0.2224 | 0.6155 | 750 | 222 | 1222 | 1.35 | | |
| 0.9 | 0.3066 | 0.6888 | 900 | 306 | 1306 | 1.45 | | |
| 1.0 | 0.3679 | 0.7310 | 1000 | 368 | 1368 | 1.5 | 1.81 | 2.45 |
| 1.5 | 0.7231 | 0.8705 | 1500 | 723 | 1723 | 1.75 | 2.41 | 4.80 |
| 2.0 | 1.1353 | 0.9366 | 2000 | 1135 | 2135 | 2.00 | 6 | |

Note that although the LO can exceed 1 for some files, the SO can *never* exceed 1. The unit list with global overflow (ULGO) is pratically efficient when the SO approaches 1, for then most cells are in use. At this time the LO is usually greater than 1. Morris and Johnson unfortunately have made their calculations using the LO defined above to produce this equation for the average number of looks, $\hat{L}_o$, for the ULGO;

$$\hat{L}_o = 1 + LO/2 \qquad (10.3.5)$$

For instance, for an LO of 2, $\hat{L}_o = 2$. Since the file is twice as large as the list, at least half the file is in overflow; yet only two looks, average, are needed to locate a record!

It is clear that, given storage S, the way it is divided into main, L, and overflow, H, affects its efficiency. Some overflow is needed because we cannot expect a perfectly randomizing mapping. Increasing the size of overflow at the expense of the main list makes up for this imperfection and keeps the number of looks low.

Table 10.3.1 demonstrates that the number of looks is significantly reduced by the overflow probe technique, especially for high list occupancy, and with only a small consequent extra storage requirement.

### Local and Intermediate Overflow

It should be clear that there is no need in this format to have local overflow that would provide *some* overflow for *each* list cell.

**Fig. 10.3.3.** Mapping with intermediate overflow

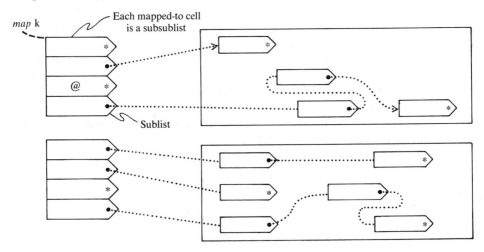

Intermediate overflow does serve a purpose, however, where information is stored on DASD media. Here mapping takes us to a cell (subsublist) in a section (sublist) of the medium. Place the sublist and its intermediate overflow on the same cylinder. Since overflow is chained only through this single cylinder of the medium, much time will be saved during probe because no seeks are required.

In Fig. 10.3.3 we see that several subsublists (cells) can have single intermediate overflow associated with them and that these can be in the same general area.

## 10.4 HASHED HEAD
## MULTIPLE LINKED LIST FILE

### Need

The file structure described here is an improvement on the unit sublist mapping with overflow and stems from a deficiency in that method. You recall that reducing the number of cells per sublist to a single cell minimizes waste for empty subfiles. But as described earlier, at least one cell per sublist is required as a starting point for any subfile, even when that subfile is empty. Thus for an empty subfile, we still have one cell of overhead —record size space that is unused.

To reduce this overhead still further, this cell too can be replaced. Instead of mapping into the first cell of the sublist, the mapping function now takes us to a head. This is one of a set of heads, pointers to linked lists. A head is indeed very small and consists simply of a pointer to a cell that contains the first record of a subfile. Now for an empty subfile, only the head is empty, and we have wasted only a small amount of space, perhaps one word. And it is really not wasted since it keeps track of the fact that this subfile is empty!

The head points to the beginning of the linked list that is variable in size according to the subfile being stored. These linked lists reside in a linked list area. A single area can be used by all the subfiles that comprise our file. We shall see later that another alternative is to partition the list area to suit the medium that stores the list. This means that a probe for a given record can stay in approximately the same physical area.

We have looked at many file structures so far, and this file could be seen in two representations that are actually equivalent:

- a hashed head list that is entered by using the mapping function and makes the choice of which of many linked lists our search will continue in;

- this is also a mapping into a neighborhood that is actually empty but points to an overflow area into which all records for the subfile are chained.

### Structure

Figure 10.4.1 shows in graph form this file structure. The key (1) of a record for search or appending is mapped by some function (2) into the head list, **R** (3). This yields a directory entry $R_i$ (4). The entry contains a pointer to the beginning of a linked list (5). Corresponding to each directory entry, $R_i$, there is a linked list, $_iL$, in the list area (6). The sublist is so named even when it is empty. Cells of the sublist are reached by extracting pointers from successive cells (7). The last cell in the sublist contains a terminal pointer, * (8).

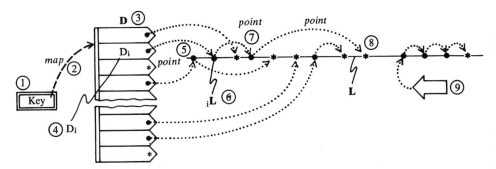

**Fig. 10.4.1.** Hashed directory multiple linked list file

One of two techniques described earlier can be employed to keep track of empty available cells for new records:

A. The space pointer (9) points to the available area in the empty portion of the list area, ideal for variable length records.
B. A space list can do the same job for fixed records where the file is subject to change.

As discussed earlier with respect to linked lists, when much deleting is done, deleted cells will get lost for the space pointer technique; whereas the space list maintains the availability of cells from which records have been deleted.

Then the list area, L, contains sublists, $_iL$, of which there are heads in the head list, **R**. Each sublist, $_iL$, contains cells that we shall label in the order in which they are linked together as $_iL_j$, where $_iL_1$, for instance, is the first cell that is pointed to by the directory.

Records appear in the cells of the linked list generally in the order in which they were added to the sublist. It is possible to require each sublist to be ordered so that the record with the smallest key of the subfile is contained in $_iL_1$, and so forth. The extra manipulation required to keep this order really buys us nothing, except when searches are frequently performed for absent records.

### Search

Let us see how we go about searching for a record with a particular key in the hashed directory multiple linked list file. As before, the first action is to convert the key, using the mapping function to find a directory entry, thus:

$$map \; k = R_i \qquad\qquad (10.4.1)$$

If the head cell is empty, this is a hit and we stop:

$$(R_i) = @: \quad stop \qquad\qquad (10.4.2)$$

Otherwise, the head points to the first record in the sublist. We obtain that pointer, go to the cell, pick out the record key, and compare it to the key sought:

$$(R_i) = point \; map \; k; \qquad ((R_i))_K: \quad k \qquad\qquad (10.4.3)$$

If this is the desired record, we stop. If not, we obtain the pointer from the cell, go to the next cell in the sublist, obtain the key of the record kept there, and compare it to the key sought:

$$((R_i))_K = k: \quad stop; \qquad (point^2 \; map \; k)_K: \quad k \qquad\qquad (10.4.4)$$

This action continues until the pointer that we extract from a sublist cell is a terminal pointer; then we must assume that the record sought is absent:

$$point^j \; map \; k = *: \quad stop \qquad\qquad (10.4.5)$$

**example**      Figure 10.4.2 presents an example of how a search is performed. The mapping function is modulus M division. To make the mapping more practical, we offset this function by the location of the head list [R]. We also scale the mapping by s, according to the size of the head:

$$map \; k = [R] + k \; (mod \; M)*s \qquad\qquad (10.4.6)$$

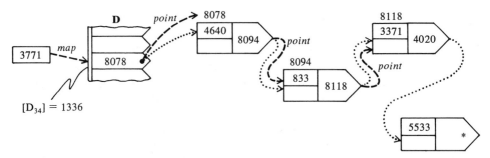

**Fig. 10.4.2.** Searching a HDMLL file

For the example, we have

- **[R]** = 1200—the head list is located at 1200;
- M = 47—this is division modulus 47;
- s = 4—the head is four bytes long.

Then (10.4.6) becomes

$$map\ k = 1200 + k\ (mod\ 47)*4 \tag{10.4.7}$$

Let us now search for a record with key 3317. Applying (10.4.7) we come up with the location of the head for this mapping:

$$map\ 3317 = 1200 + 34 * 4 = 1336 = R_{34} \tag{10.4.8}$$

Notice from the figure that the head is not empty. It contains a pointer to cell 8078, which in turn contains a record with key 4640:

$$point \cdot map\ 3317 = 8078; \quad (8078)_K = 4640 \tag{10.4.9}$$

Search proceeds as noted until we find the record:

$$(point^3\ map\ 3317) = (8118)_K = 3317 \tag{10.4.10}$$

Let us now search for a record with key 2666. When we apply (10.4.7), we come up with the same directory entry:

$$map\ 2666 = 1336 = R_{34} \tag{10.4.11}$$

We follow the trail of pointers until we find a terminal pointer in the cell at 9020:

$$point^5\ map\ 2666 = *: \quad stop \tag{10.4.12}$$

This record is absent.

### Appending

To append a new record, go through the search procedure as described above. After determining that the new record is truly absent from the subfile, obtain space from the space pointer, insert the record in this cell, and attach it to the end of the sublist, adjusting pointers as required.

*example*    Figure 10.4.3 presents an example. We wish to append a record with key 2661. The search is conducted as shown in Fig. 10.4.2. In Fig. 10.4.3 we see the mapping (1) and how we end up at the last cell of the sublist (2). Next we go to the space pointer and pick up the

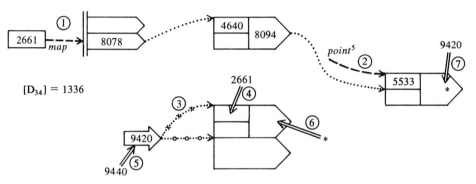

**Fig. 10.4.3.** Appending to a HDMLL file

address of the next available space in the list area (3). The record is now placed here (4). Assuming that the record is 20 bytes long, including its pointer, update the space pointer to point it to the next available space (5). The terminal pointer is required in the new cell (6). Finally, the last cell that we found in the sublist is set to point to this cell (7).

### Deletion

To delete records, the techniques developed in Section 7.4 are applicable. It is left to the reader to do an example for his own elucidation.

### Local or Intermediate Overflow

As presented above, the list area is one contiguous set of cells. A more convenient way to store large lists is to keep each subfile in a different area of an auxiliary medium. If each subfile requires a number of tracks, for instance, then search time can be decreased by allocating a cylinder to each sublist. Available space in the cylinder is monitored by a space pointer. This is shown in Fig. 10.4.4.

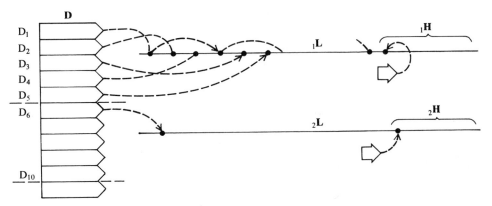

**Fig. 10.4.4.** An HDMLL file with intermediate overflow

Now the head list determines the location of the first cell in the linked sublist. The entire sublist is found in the same cylinder so we need not travel back and forth to reach successive cells. If enough space is available, in fact, the entire cylinder may be transferred to CORE and search performed there.

For smaller sublists, it is clear that when several sublists occupy the same area, intermediate overflow can provide space for this *set* of sublists.

### Analysis

The structure of this file is almost identical with that discussed in Section 10.3, so that most of the results pertain. One extra look is always required, however, since the directory does not contain a record but rather a pointer to a record.

But this look is not the same kind of look that is required for the hashed mapping with overflow. Each look within a sublist requires comparisons

- to check the record key for equal;
- to extract the pointer;
- to travel to a different area on the storage medium.

The look at a head for this file structure only obtains a pointer and is a much simpler action.

**PROBLEMS**

10.1 Consider the file **f**, a list **L** associated with this file, and overflow, **H**, for **L**. Explain the relation among them.

10.2 A list **L** consists of sublists and any sublist $L_i$ consists of subsublists, one of which is $L_{ij}$. Explain local, intermediate, and global overflow and its relation to **L**, $L_i$, and $L_{ij}$.

10.3 A semidense sublist $L_i$ can be considered as two sublists, $L_i'$, and $H_i$. Explain.

10.4 Explain the relation between a subfile $f_i$ (or logical sublist) and the associated sublist, $L_i$, and its overflow, $H_i$.

10.5 What means are there to delineate for local overflow
(a) the beginning of an overflow area $[H_i]$?
(b) the end of *occupied* overflow?
(c) the beginning of empty overflow?
(d) the end of empty overflow?

10.6 Repeat Problem 10.5 for intermediate overflow.

10.7 Repeat Problem 10.5 for global overflow.

10.8 Show in colored digraph form a portion of a double directory file with intermediate *and* global overflow
(a) before intermediate overflow is exhausted;
(b) after intermediate overflow is exhausted.

10.9 Contrast and compare mapping with global overflow for a
(a) unit sublist,
(b) multicell sublist.

10.10 For mapping a unit sublist with global overflow (MUSGO), having ordered sublists, explain with an example each of the following:
(a) search, record present;
(b) search, record absent;
(c) appending;
(d) deleting.
Be sure to consider *all* alternatives—hit in sublist, hit in overflow, and hit for terminal record in subfile.

10.11 For MUSGO with unordered sublists, repeat Problem 10.10.

10.12 Consider the twin entry multilevel directory file (TEDF), with only two entries in the directory, subdirectory, subsublist, etc. Explain
(a) how overflow could be provided to append new records;
(b) how this would affect the directories;
(c) what you say about a file with any changes occurring.

10.13 For the hashed directory multiple linked list file (HDMLLF):
    (a) Contrast with the MUSGO for
      • space occupied,
      • search,
      • appending and deleting.
    (b) Explain search, append, and delete.

10.14 Consider a HDMLLF with directory **D** and list **L** (although all the list is in overflow), where **L** is on a DASD. It would be beneficial to partition **L** into sublists $L_i$, i = 1 to m.
    (a) Why?
    (b) How would you decide how large to make m?
    (c) How many subfiles would be in each sublist and why?
    (d) Where would each sublist reside on the DASD?
    (e) During a search,
      • where is **D**?
      • where is $L_i$?
      • how does each get there?

In the following problems, use Table 8.P.1.

10.15 Let n = 16 so that

$$num\ \mathbf{M} = 256; \qquad num\ \mathbf{M}_i = 16$$

where $\mathbf{M}_i$ is semidense and $\mathbf{D}_M$ is the directory for the SLDF called **M**. Let [**M**] = 75,000 and [$\mathbf{D}_M$] = 110,000. Determine the boundaries and contents of $\mathbf{M}_1$ through $\mathbf{M}_{16}$ for **M**. Determine the boundaries and contents of the directory $\mathbf{D}_M$.

10.16 Show how to search for the records in Table 10.P.1.

10.17 Append (+) or delete (−) the records in Tables 10.P.2 and 10.P.3 to show the effect on the $\mathbf{M}_i$ concerned and on $\mathbf{D}_{mi}$ when a change is required.

10.18 Create **I** from Table 8.P.1, a two-level directory file with intermediate overflow $\mathbf{H}_I$, $num\ \mathbf{H}_I = 36$, and consisting of six sublists. For the directory $\mathbf{D}_I$ and subdirectories $\mathbf{S}_{Ij}$, we have

$$num\ \mathbf{I}\ \ = 256; \qquad num\ \mathbf{I}_i\ \ = 36; \qquad num\ \mathbf{I}_{ij} = 6$$

$$num\ \mathbf{S}_I\ = 36; \qquad num\ \mathbf{S}_{ij}\ = 6$$

$$num\ \mathbf{D}_I = 36$$

$$num\ \mathbf{H}_I = 36; \qquad num\ \mathbf{H}_{I_i} = 6$$

In the following problems, use the problem file **f** in Table 10.P.1, the change file **c** in Table 10.P.2, and the transaction file **t** in Table 10.P.3. Each record contains four fields in this order:

**Table 10.P.1.** Problem File **f**

| Telephone | Name | Address | Town |
|-----------|------|---------|------|
| 8977754 | Barke John P | 30 Roosvlt | Hmpsted |
| 5283378 | Serif Medina | 6 Ann Ct | Freport |
| 2238026 | Welcome Wag Inn | 1294 N Grand Av | Baldwin |
| 3450025 | Welwood Cemtry | Welwd Av | Pinelawn |
| 4454773 | Annenberg Marvin | 187 Brit La | Hillsvl |
| 2844225 | Lucarelli Orlando | 47 Davsn Av | Ocnsid |
| 6161067 | A&A Catch Basin Cleaners | 174 Jericho Tpk | Flrl Pk |
| 7829990 | Bennett Eileen McG | 190 1st St | Minola |
| 2134502 | Nascimento Tony | 219 Horton Hwy | Minola |
| 3370336 | Lopez Richard Jr | 47 Grafton | GrdnCtyPk |
| 2020068 | Weber Mary Mrs | 46 Smith | GlnCov |
| 2584457 | Hong Kong Men's Shop | 173 Sunrise Hwy | PkvlCntr |
| 6699600 | Combes Elnathan | 218 Ct Hse Rd | FrnklnSq |
| 4507287 | Sutliffe R V Mrs | 15 Buck La | Levitwn |

**Table 10.P.2.** Change File **c**

| Telephone | | Name | Address | Town |
|-----------|---|------|---------|------|
| 3991688 | + | Danzato Philip N | 1143 Hmpstd Tpk | FrnklnSq |
| | − | Combes | | |
| | − | Weber | | |
| 6729799 | + | Rea Stephen W | 1510 Terrace Blvd | NwHydePk |
| 5950180 | + | McMahon Michael E | 50 Meadow | GrdnCty |
| | − | Nascimento | | |
| 2871599 | + | Gaba Richard M | 470 Laurl Rd | RkvlCtr |
| 2245902 | + | Bell Roger J | 144 Washington Av | VlyStrm |
| 8510832 | + | Zambuto Jos A | 584 Adams Av | WHmpstd |

**Table 10.P.3.** Transaction File **t**

| | | |
|--|--|--|
| 2584457 | 8977754 | 3370336 |
| 7829990 | 2245902 | 2020668 |
| 6729799 | 6699600 | |

1. key, fixed on seven characters;
2. name, variable, up to twenty-eight characters;
3. address, variable, up to eighteen characters;
4. town, variable, up to nine characters.

Mapping is done by division mod 101.

10.19 Enter **f** into a list consisting of a main list, **A**, and global overflow, $H_A$.

$$[A] = 30{,}000; \quad num \ A = 101$$
$$num \ A_i = 4; \quad len \ A_{i1} = 60; \quad len \ A_{i4} = 65$$
$$[H_A] = 55{,}000; \quad len \ H_A = 12{,}000$$

10.20 Show how to find in **A** the records identified in **t**.

10.21 Use a space stack, $S_A$, to keep track of space in $H_A$ and make changes from **c**.

10.22 Enter **f** into a list consisting of the main list, **B**, with overflow $H_B$ where

$$[B] = 30{,}000; \quad num \ B = 101; \quad len \ B = 6565$$
$$num \ B_i = 1; \quad len \ B_i = 65$$
$$[H_B] = 40{,}000; \quad len \ H_B = 13{,}000$$

10.23 Repeat Problem 10.20 for **B**.

10.24 Use $S_B$ for space in $H_B$ and make changes from **c**.

10.25 Enter **f** into a list of 101 cells of length 65 characters in **D** with overflow of variable length in overflow $H_D$ where

$$[D] = 30{,}000; \quad num \ D = 100; \quad len \ D = 6565$$
$$num \ D_i = 1; \quad D_i = D_i; \quad len \ D_i = 65$$
$$[H_D] = 4000; \quad len \ H_D = 12{,}000$$

10.26 Repeat Problem 10.20 for **D**.

10.27 Use $P_D$ for a space pointer for $H_D$ and enter changes from **c**.

10.28 Enter **f** into **E**, which consists of two parts:
   (a) A directory $D_E$ of 101 entries of 5 characters each to point to the first record in $H_E$.
   (b) All records are in $H_E$.
   (c) Each record is of variable length of up to 65 characters; the last 5 are a pointer to the next record in that subfile.
   (d) $P_E$ is a pointer to the next empty space in $H_E$.
   We have

$$[D_E] = 30{,}000; \quad [H_E] = 31{,}000$$

10.29 Show how to find the records identified in **t** of Table 10.P.3.

10.30 Make the changes found in **c** using $P_E$ of Table 10.P.2.

# CHAPTER 11

# Tree Representation

## 11.1 BINARY TREES

### Definition and Need

The linked list is particularly attractive for appending and deleting records, but it does not permit us to do binary searches. We have seen in Chapter 5 and others that follow how truly effective the binary search can be in reducing the number of looks required to find the record. Can we have the advantages of both? The binary tree provides us with both advantages, and at only a small price.

The **binary tree** is a bifurcating arborescence:

- Each node is a data node.
- Each node contains two forward pointers.

A variant of the binary tree keeps data only at the leaves. This requires us to traverse the tree completely to find the data and is hence less efficient.

### Binary Search

Figure 11.1.1 shows an ordered list of fifteen items and the path followed to do a binary search. Enter the list at hank. Compare the key sought with hank and take one of these actions:

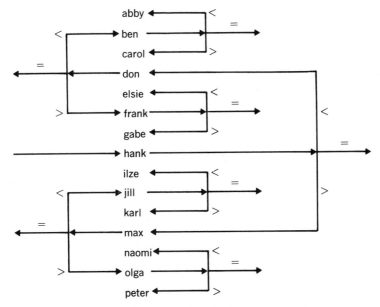

**Fig. 11.1.1.** A binary search applied to an ordered list

- For *equal*, the record is found.
- For *low*, take the upper branch in the figure.
- For *high*, take the lower branch.

We have seen how the binary search determines which way to go and then finds the fence cell for the next look. Calculations have to be made at each step of the way. Why not precalculate where to go and provide a pointer for each alternative?

### Binary Search Tree

The binary search tree for the ordered list of Fig. 11.1.1 appears as Fig. 11.1.2. Not only is this a tree but it is a linked list where each cell contains two pointers, one for each alternative search path.

Each node is a cell, and the cell contains a record. One alternative for the node cell is shown in Fig. 11.1.3. Here the cell G contains a record with key g. Other fields of the record are shown in the figure. At the right we see a high pointer $G_H$. This pointer is a link to the next cell to be examined (H) should our comparison reveal that we should take the "high" path. A low pointer $G_L$ (F) is provided should the decision be to take the "low" path. For *equal*, the record is found—go no further.

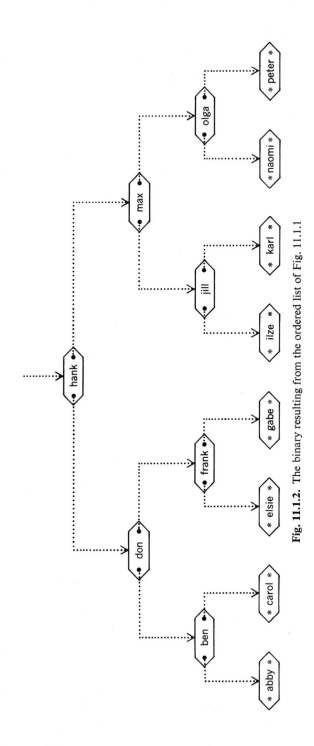

**Fig. 11.1.2.** The binary resulting from the ordered list of Fig. 11.1.1

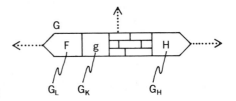

**Fig. 11.1.3.** A binary tree data cell

We use subscripts to distinguish the two pointers and the key:

- The key field has the subscript K $((G_K) = g)$.
- L is for lower pointer $((G_L) = F)$.
- H distinguishes the high pointer $((G_H) = H)$.

Another alternative is to keep the datum in a different list. Then each cell is missing the fields that comprise the record but has an additional pointer for which the subscript D (data) is used; this points to the location where the record is found. Figure 11.1.4 shows a binary tree cell using the data pointer format.

The file using a data pointer cell as in Fig. 11.1.4 might be called a **binary tree total directory file**. It has the advantages of both the linked structure for appending and deleting records and the ordered list structure for performing binary searches. Additionally it can accommodate variable length records.

An example of such a file is presented as Fig. 11.1.5.

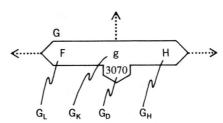

**Fig. 11.1.4.** A binary tree data pointer cell

## 11.2 USING BINARY TREES

### Searching

It should be fairly obvious how searching is done. As we reach any cell, we compare the key sought with the key contained there. *Equality* indicates that the record is found. If the cell uses a data pointer, that is extracted to reach the record.

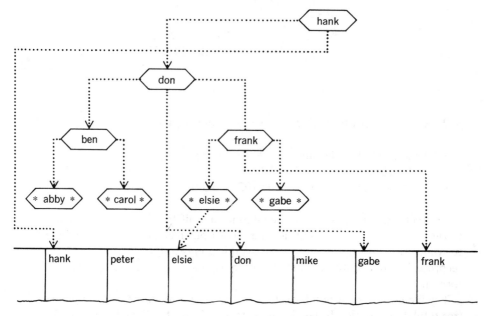

**Fig. 11.1.5.** A binary tree total directory file

For *inequality* the high or low pointer is taken, according to the result of the comparison. Search continues until the pointer field chosen as a result of the comparison contains a terminal indicator (*). Then it is clear that the record is absent.

### Append

To append a new record, we *search equal greater* until we reach a terminal pointer, taking the proper forks in the road until we reach this dead end. This terminal pointer will be replaced by a pointer to the new record when we add it.

As with other linked lists, we should have a space list of available cells. A cell is obtained and a new record (or data pointer) placed there. The pointer in the preceding record is set to point to the new cell. Both the high and low pointers in the new cell have terminal indications.

In Fig. 11.2.1 jack is to be added to the tree of Fig. 11.1.2. We follow the trail to ilze, who is a leaf. After a cell is obtained for jack and he is put therein, we set the high pointer of ilze to point to jack.

In Fig. 11.2.2 we see how henry and irv are later added to the tree.

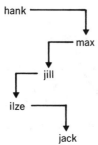

**Fig. 11.2.1.** Appending jack

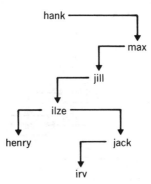

**Fig. 11.2.2.** Appending henry, then
irv after jack has been appended

### Delete

It is easy to delete a leaf. First, follow the path to the leaf;
then return to the predecessor of the leaf and insert a terminal pointer to show
that the leaf has been cut off. The leaf cell may be returned to the space list,
if one is kept. Figure 11.2.3 shows how elsie is deleted.

When the record to be deleted is not a leaf and is either the root or a rootlike node, we move some selected leaf to the position occupied by the deleted
cell. The question is which leaf to use. Figure 11.2.4 shows a portion of a tree
where max is to be deleted. The figure indicates that either karl or naomi may
take max's place. How so? karl is the largest key in the low branch, and

**Fig. 11.2.3.** Deleting elsie

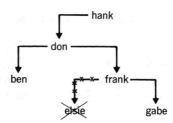

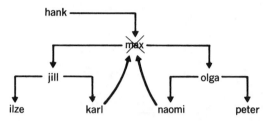

**Fig. 11.2.4.** To delete max, he may be replaced by either karl or naomi.

naomi is the smallest key in the high branch. Either one of these may replace the deleted record. (Why?)

Figure 11.2.5 shows the tree of Fig. 11.2.4 after max has been replaced by karl. Figure 11.2.6 shows how the pointers in karl and hank are reset to reflect the change. The cell occupied by max may be returned to the space list.

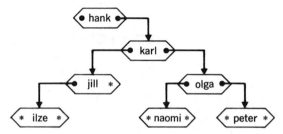

**Fig. 11.2.5.** The tree after max has been replaced by karl

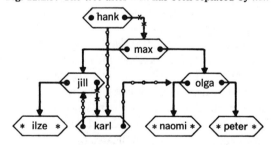

**Fig. 11.2.6.** Pointer changes required to replace max by karl

### Doubly Linked Binary Search Trees

During search we often retrace our steps. A reverse pointer in a binary tree cell would be helpful in this respect. Figure 11.2.7 shows how such a reverse pointer is used to point to the cell previous to this one on the next lower level (toward the root). The subscript R indicates the reverse pointer.

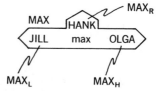

**Fig. 11.2.7.** A binary tree data cell
with a backward pointer

An additional pointer means something extra to update. As always, we
weigh the advantages against the cost.

### Separate Subtrees

As with the multiple directory system, the binary tree file
lends itself to partitioning. In Fig. 11.2.8 we see how a very large tree can be
broken into subtrees. The root subtree contains the root and might be kept
resident in MEMORY. For a disk file, the intermediate subtrees might be kept
on each disk pack. Finally, there might be one leaf subtree containing the
records proper in each cylinder.

**Fig. 11.2.8.** A large tree may be broken up into multiple subtrees, as
shown.

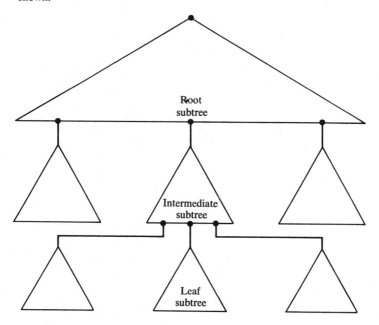

## 11.3 BALANCED TREE CREATION
## AND ALTERATION

There are many ways in which a binary tree can be created. We shall only discuss three here:

- binary search of a linear list;
- sequential entry into a prefabricated tree;
- random entry into a prefabricated tree.

### Creating by Binary Search

The purpose of the binary tree is to reproduce the binary search procedure without employing the algorithm. The tree may be created within the list proper by designating two extra fields in each record. Or, the tree can be created elsewhere in a linked list, using cells withdrawn from a pool or a space list and entering the proper pointers. Since the former method is more economical in both time and space, only it will be examined.

Suppose we start with the ordered list of Fig. 11.1.1 examined earlier to illustrate how a binary search is converted into a binary tree. Assume that cells in the list contain room for two additional pointers, a *low* and *high* pointer. To construct the list, shown in Fig. 11.3.1, the binary search algo-

**Fig. 11.3.1.** Converting an ordered list into a binary tree using the binary search algorithm, where space for high and low pointers has been provided.

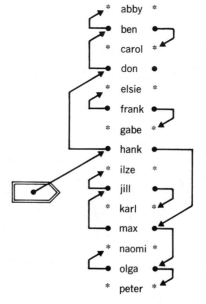

rithm is applied to the list. Once the entry point hank is determined, his location is entered into the head. Thus hank is contained in the first fence cell.

When the binary search algorithm is applied to find the second fence, its action depends on the comparison of the search key with hank. To construct the tree we consider both cases and find both fences. The fence locations are entered into hank's *low* and *high* pointer.

Next, proceed to these fence cells to enter pointers into them. Use the algorithm and find two pointers for each record.

Eventually the algorithm tells us that cells are leaf cells and to put terminal indicators in both pointer fields.

The resulting tree appears in Fig. 11.3.1. When you look at it, the question arises, why bother with all the pointers when we have an ordered list for which we can apply a binary search? Only when the tree is altered does its usefulness become immediately apparent. We shall soon see how the tree is altered. But to summarize, the new record is placed at the end of the list where it would certainly be out of order if this were not also a tree. Then the tree is updated to include the new record. Since this is done simply by altering links, the original tree is in no way disturbed.

### Bottom up Construction

To determine the form of the optimum binary tree we need only know the number of records that it will hold. Consequently, if we have an incoming list that has been ordered and is on such medium as tape or disk, the records can be routed to the proper node and the tree can be filled in on the fly. Thus in Fig. 11.3.2 we see such a tree in the midst of filling. T1 is the

**Fig. 11.3.2.** The tree may be prefabricated and then records entered in the proper order.

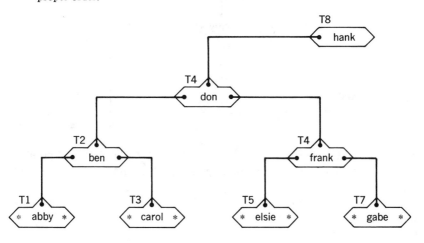

cell allocated to the lowest record, and it is a leaf on the extreme left. The first record coming in, **abby**, is entered there. T2 is allocated to the next record, and **ben** will go there, and so forth.

We need only know the size of the tree to prefabricate the structure and fill in the records as they arrive.

### Appending

Not only is appending a necessity for maintaining binary trees, but also it is a method for building them, as we shall examine later in the section.

To append a record, first place it in an empty cell. Then do a search for the record along the binary tree, taking lower or higher branches according to the direction of the inequality between the new record and the one encountered. If the record is absent, the path takes us to a leaf. A leaf is distinguished by two terminal pointers. Choose the high or low pointer according to whether the new record key is greater or less than this leaf cell. Fill this pointer with the address of the new record.

Figure 11.3.3 shows how several new records are added to our binary tree. First to be appended is jack. We travel through the tree, passing hank, max,

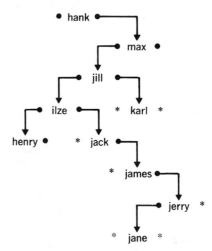

**Fig. 11.3.3.** Adding new records in this order: jack, henry, james, jerry, jane

and jill, to reach ilze. jack is larger than ilze, so the high terminal pointer in ilze is pointed to the cell containing jack, which now has two terminal pointers.

The figure shows how james, henry, jerry, and jane are appended in that sequence.

As you can see, the addition of only a few records will make our tree way out of balance. That is, the number of looks required to find jane (8) is much more than required to find *any* of the other records of the original tree, such as abby (4), karl (4), and don (2).

It is interesting to note that this length depends on the order in which the records are presented for appending. For instance, the same records were added to the original tree but in the order jane, henry, jack, jerry, james, to produce the result in Fig. 11.3.4. Now it only takes five looks to find jane and six for jack or jerry. The tree of which part is presented in Fig. 11.3.4 is more attractive from this point of view.

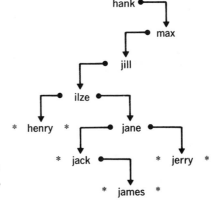

**Fig. 11.3.4.** Adding the same records in a different order: jane, henry, jack, jerry, james

### Deleting

To delete a *leaf* such as ilze, we merely enter a terminal pointer in the cell that points to ilze, namely JILL. To delete a branching node it is almost as simple. Figure 11.3.5 shows the action required when we wish to delete max, who is in the midst of things. The method is to find an appropriate leaf and simply move the contents of the leaf cell into the cell that now contains max. There are two optimum choices:

- the record with the highest key in the low branch (karl);
- the record with the lowest key in the high branch (naomi).

You can work out *why* this choice is optimum. To put this intuitively, karl is higher than jill or any other record pointed down from max. Therefore karl can replace max and point downward. Since karl is smaller than max, he too will point upward to olga.

Figure 11.3.6 shows the right branch of the tree after karl has replaced max.

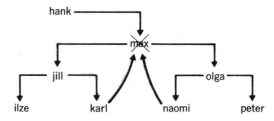

**Fig. 11.3.5.** Deleting max

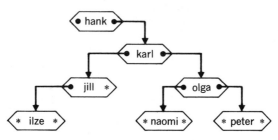

**Fig. 11.3.6.** The tree with max deleted

### Random Construction

The third way of constructing balanced trees is to suppose that records are presented in random order and hope that this is true. As each record is examined, it is appended to the existing tree by the action described earlier. No attempt is made to anticipate which records will arrive. As long as the arrivals are truly random, the tree ends up fairly balanced, as explained in Section 11.4.

## 11.4 BALANCED TREES

### Comparing Binary Trees

We have seen in Section 11.3 that when trees grow like Topsy, they are not necessarily the best ones for all searches. How can we compare one tree with another as far as searches are concerned?

We define the **height of a branch** as the number of nodes between the root and the leaf, inclusively. The height of a branch tells us how many cells must be looked at before we reach a leaf and have exhausted all the possibilities for finding a particular record. Intuitively, you can see that if all the branches of our tree have approximately the same height, the tree is better for searching than a similar tree containing the same records but so distributed that some branches are much longer than others.

What is the best we can do? Well, if we have exactly $2^h - 1$ nodes, then we can construct a tree where every branch has height of h. For instance, with our tree in the beginning of the chapter we had fifteen records. $2^4$ is 16, and we saw that we could design a tree using the binary search algorithm so that every branch had a height of 4.

Things don't usually work out so well. For instance, if we had only fourteen records, one of the branches would have to be of length 3, but we would still have an optimum tree. We define an **optimum tree** then as a tree for which all branches have a height of either h or h − 1. There are only two lengths of branches. An optimum tree is hard to achieve and especially hard to maintain.

A compromise is the balanced tree. To define it, first we define the **height of a tree**, which is the maximum height of any of its branches. Now, given a tree, pick any node. From this node two subtrees are growing. Find the height of each subtree. The height of each subtree must not differ by more than one to have a balanced tree. To reiterate, a **binary tree** is called **balanced** if the height of the upper subtree at every node never differs by more than one from the height of the lower subtree. A balanced tree is presented in Fig. 11.4.1.

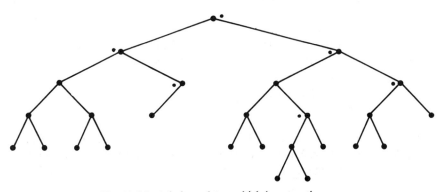

**Fig. 11.4.1.** A balanced tree which is not optimum

Such trees put a reasonable upper limit on the length of searches we might make.

We cannot expect the result of random construction to be either an optimum or a balanced tree. But, on the average, it has been shown that random construction is generally in balance or not much off balance. The problem is that data do not always arrive in a random sequence. For instance, according to the sequence of arrival indicated by the circled numbers, the unbalanced tree of Fig. 11.4.2 will be constructed. This is poor and we should be able to exercise some control over the construction.

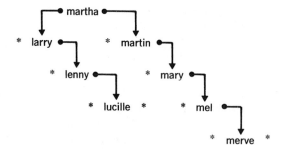

**Fig. 11.4.2.** A very unbalanced tree resulting from the order in which records were presented

### Approach to Balancing

Let us assume that we have, at the moment, a balanced tree and a new record that must be appended to it. This new record may throw the tree out of balance. At that moment we do something with the tree to rebalance it. *To restate this, each time we add a new record, rebalance the tree.*

To simplify the rebalance procedure, require each node to carry balance information. Recall that any node of a balanced tree can be in three conditions:

- **completely balanced,** so that the height of both subtrees is equal;
- **high balanced,** so that the high subtree is one node longer than the low subtree;
- **low balanced,** where the low subtree is one node longer than the high subtree.

Otherwise the tree is **unbalanced,** when one subtree is more than one node longer than the other. Each node cell records for *both* branches whether each is fully balanced, high, or low. Two bits do this:

$$00 = \text{balanced}; \qquad 01 = \text{high balanced};$$
$$10 = \text{low balanced}; \qquad 11 = \text{error}$$

In figures that follow, a dot beside a node indicates the direction of imbalance, if any. Figure 11.4.3 gives these specific examples:

- A dot on the right indicates that the high subtree is longer (B).
- A dot to the left indicates that the low subtree is longer (C).
- No dot indicates that the node is in balance (A and D).

Figure 11.4.3 should clarify this.

**Fig. 11.4.3.** A dot at either side of a node indicates the direction of imbalance of the subtrees from it.

To add a record, always add a new leaf. In many cases, statistically approximately half, the tree remains in balance. There are three cases:

(A) Obviously if it is in total balance now, the imbalance created by the new node is acceptable.
(B) If it is unbalanced in the opposite direction, the new record balances it.
(C) It is only when the new cell is added to the out-of-balance (heavy) branch that we need remedy the situation.

All instances of (C) break down into two subcases or their variations. We examine these two cases and remedies for them below.

### Case IH, Explanation

Examine the first case where the imbalance is on the high side, presented in Fig. 11.4.4; call it IH. We see a number of nodes, S, T, U, V, and W. The *pointers* from each node are dotted arcs; *greater than*, such as $T > V$, is shown by the light solid arc (T to V). We have the following:

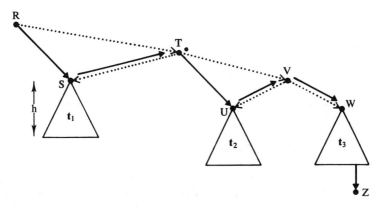

**Fig. 11.4.4.** Case IH where a new node, Z, throws a balanced tree out of balance

- Hanging from T is the low subtree $t_1$, of height h beginning with S.
- Hanging from V we find the low subtree $t_2$ of height h beginning with U.
- Hanging from V we find the high subtree $t_3$ of height h beginning with W.

It is desired to add a new node, Z, to one of the leaves of $t_3$. (The symmetric case would be where we add Z to the lower tree, $t_2$, and produces a symmetric solution, which is left to the reader.) The addition of the new leaf makes the tree unbalanced at the node T because the length of the low subtree of T is h + 1, while the length of the high subtree is h + 3.

Before we solve this case, we define the focal node (T) along the path from the leaf, the new cell, backward to the root. The first rootward node that is *presently* out of balance in the direction of the added leaf is the **focal node**. This is determined from the balance tags in the nodes we pass.

In Fig. 11.4.4 the relation of the keys at the nodes is important to realize. In contrast to the high (right) and low (left) pointer in dotted arcs, key relations shown by the light arrows show that

$$R < S < T < U < V < W \qquad (11.4.1)$$

Notice that the focal node, T, is near the left of this set of inequalities. The new imbalance has weighted us to the right and we find that the solution is to shuffle the nodes so that the high subtree pointed to by R no longer has the root T but now has the root V, as shown in Fig. 11.4.5.

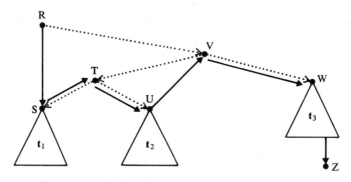

**Fig. 11.4.5.** Solution to Case IH presented in Fig. 11.4.4

### Solution—Case IH

Figure 11.4.4 presented the subtrees and the relation among the nodes that are to be readjusted. Figure 11.4.5 shows the alterations required to provide the solution. Notice that now R points to the node V. The relation among all the nodes, as indicated by the light arcs, remains unchanged, however. Now the heights from V to $t_1$, from V to $t_2$, and from V to Z are all $h + 2$. Thus this node V is balanced since the height of the low subtree is the same as the height of the high subtree from V. So is the node T.

***pointers*** Figure 11.4.6 shows the adjustments made to rebalance the tree. First, point $t_3$ to Z, which creates the imbalance at T. Then, in Fig. 11.4.6, the adjustment of three pointers is made:

1. The high pointer in R is changed from T (1) to V (2).
2. The high pointer in T is changed from V (3) to U (4).
3. The low pointer in V is changed from U (5) to T (6).

All are consonant with the inequality (11.4.1) above. The result is the arrangement in Fig. 11.4.5.

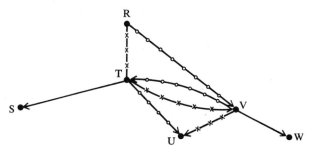

**Fig. 11.4.6.** Pointer adjustments required for the solution presented in Fig. 11.4.5

**Example**

Figure 11.4.7 portrays an example of case IH. To simplify the example, the subtrees $t_1$, $t_2$, and $t_3$ have been reduced to unit or nodal subtrees, so that actually S, U, and W are leaves containing, respectively, sid, ursala, and will. We wish to attach zelda in cell Z to W, holding will. The dot at the left of cell T indicates that this node is presently high balanced. The addition of zelda will make the tree unbalanced. Examination of the figure reveals that T is the focal node.

When the subtree of Fig. 11.4.7 is rewritten in linear form, we see that ralph points to tom; when zelda is added, the scales are tipped too much to the left, as shown in Fig. 11.4.8.

Let us readjust the pointers as required by Fig. 11.4.6 to reach the solution presented in Fig. 11.4.5, which, for the example, is then shown in Fig. 11.4.9. The high subtree is now in balance; all nodes except W are also in balance. Since W has no low pointer, a dot appears to the right of will in Fig. 11.4.9.

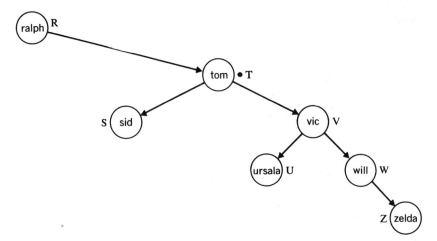

**Fig. 11.4.7.** A sample subtree requiring rebalance

**Fig. 11.4.8.** Rebalance amounts to shifting the contents of the focal node.

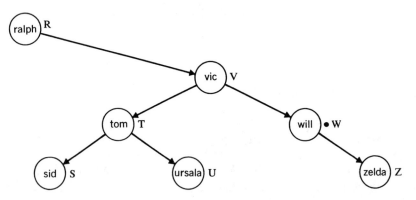

**Fig. 11.4 9.** The subtree of Fig. 11.4.7 rebalanced

## Explanation of Case IIH

Case IH, presented in Fig. 11.4.10, involves eight vertices and four subtrees. The subtrees are not of equal height: $t_2$ and $t_3$ are each of height h; $t_1$ and $t_4$ are each of height $h + 1$ and we have

$$R < S < T < U < V < W < X < Y \tag{11.4.2}$$

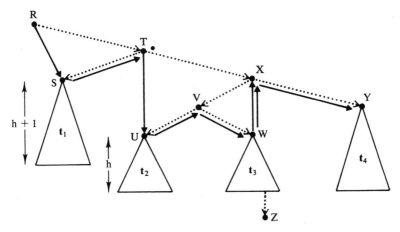

**Fig. 11.4.10.** Case IIH where addition of the node Z throws the tree out of balance

It is obvious that there is an imbalance at the node T, for we have

$$height\ Tt_1 = h + 2 \tag{11.4.3}$$

It is now required to add node Z to $t_3$. Obviously, a symmetry exists to the solution (we could add to $t_2$). A further symmetry exists around T so that if we reverse the subtrees from T, we have another variation of case IIH. The reader should examine the other three instances described.

To alter the forthcoming imbalance we wish to reorganize the high subtree of R so that V becomes now the root of that subtree instead of T.

### Solution

Figure 11.4.11 shows the solution for case IIH. Dotted lines are for pointers. Again the light lines convey the *greater than* relation

**Fig. 11.4.11.** The rebalancing solution for Fig. 11.4.10

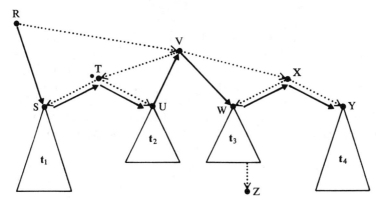

among the record keys involved in the alteration of the subtree. These relations are maintained. The subtree is now in balance; only one node, T, is out of balance. Since the relations existing within the subtrees $t_1$, $t_2$, $t_3$, and $t_4$ have not altered, they need not be examined.

***pointers***     The solution requires alteration of five pointers. The action required is

  (a) $V \longrightarrow R_H$

  (b) $T \longrightarrow V_L$

  (c) $X \longrightarrow V_H$                                           (11.4.4)

  (d) $W \longrightarrow X_L$

  (e) $U \longrightarrow T_H$

This is shown pictorially in Fig. 11.4.12.

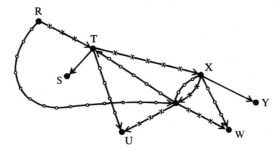

**Fig. 11.4.12.** Pointer adjustments to implement the solution of Fig. 11.4.11

### Example

Figure 11.4.13 presents an example where the records have numerical keys. Notice that $t_2$ and $t_3$ are the null subtrees (nodes) U and W to simplify the example; however, $t_1$ and $t_4$ must have height one greater than $t_2$ or $t_3$. Hence we see that $t_1$ consists of S and the leaves S1 and S2 and that $t_4$ consists of Y and the two leaves Y1 and Y2. To the subtree from T we wish to add a new node with key 72, which we call Z1. Other alternatives, Z2, Z3, and Z4 containing, respectively, records with keys 75, 77, and 79, are also illustrated in the figure and in the solution that follows in Fig. 11.4.14. These are shaded to indicate they are different problems. The addition of one new record Z1 results in only one low balanced node, namely U, after balancing. Notice that any of the four new (or all) records could be added using the same balancing relative to the focal node T.

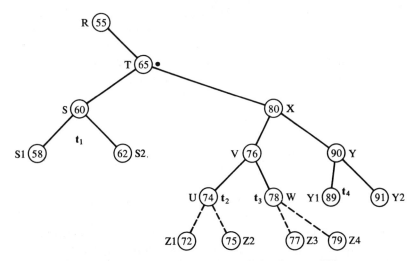

**Fig. 11.4.13.** A sample subtree for rebalancing case IIH

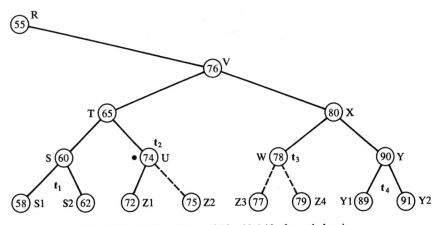

**Fig. 11.4.14.** The subtree of Fig. 11.4.13 after rebalancing

**Action**

In practice the routine that appends a new record performs a search first. As it examines each record, it checks the balance bit, possibly using a stack to maintain the information for use when appending is actually done. Search continues until the leaf is found where appending will be done. The stack is now popped to find the focal vertex and to note which case has arisen. If appending does not require rebalance, check to see if it does affect the balance at one or more nodes. The balance bit for these must be readjusted to keep the record straight for the next time appending is done.

If rebalance is required, it is done as described. The stack indicates the focal node that is used in the rebalance process. After rebalance, the balance bits at affected nodes are reset for the next append action.

## 11.5 MULTIWAY TREES

### Why ?

The multiway tree allows us to make a multiway decision at each node. This reduces the number of levels in a balanced tree and consequently means that to find a desired record we visit fewer nodes and consequently fewer cells in the tree. Thus the multiway tree can decrease the number of looks required to find a record.

Another advantage accrues when each level represents a different storage area, such as a different cylinder in a Direct Access Storage Device. Then going from one node to another requires DEVICE time, which is much more valuable and lengthy than COMPUTER time.

How many ways do we wish to go from a single node? The next step from a binary node is the ternary node, which we examine next.

### The Ternary Tree

Figure 11.5.1 illustrates a **ternary tree** where each node is capable of branching in three directions.

**Fig. 11.5.1.** A ternary tree

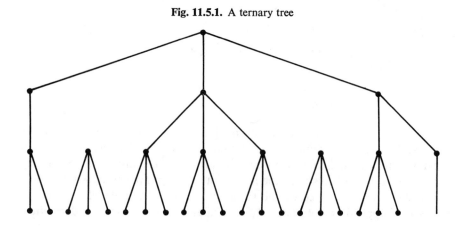

Figure 11.5.2 illustrates one implementation of the ternary node. In the figure we see a node called M. It contains three pointers and two records. The records, of course, can be replaced by data pointers. Subscripts distinguish the five subfields of the cell:

- The low, medium, and high pointers are called $M_L$, $M_M$, and $M_H$, respectively.
- The data fields are double subscripted to distinguish the low and high datum, as $M_{DL}$ and $M_{DH}$, respectively.

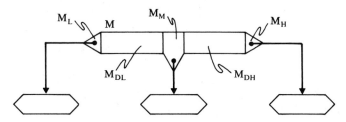

**Fig. 11.5.2.** Data cell for the node of a ternary tree

Now what is the relation among the pointers and the keys of the records contained in the node? It should be clear that

$$(M_L) < (M_{DLK}) < (M_M) < (M_{DHK}) < (M_H) \qquad (11.5.1)$$

Restated:

- The low pointer points to a branch wherein all the records have keys lower than that of $M_{DL}$.
- The middle pointer, $M_M$, points to a branch that contains records, the keys of which lie between the keys of the two records found at this node, $M_{DL}$ and $M_{DH}$.
- The high pointer points to a branch, all of whose records have keys higher than the node record, $M_{DH}$.

To do a search, first look at the key of $M_{DL}$:

- For *low*, follow $M_L$.
- For *equal*, we have the record.
- For *high*, examine the high record, $M_{DH}$.

For the high record, perform a similar set of actions:

- For *low*, follow the middle pointer, $M_M$.
- For *equal*, we have the record.
- For *high*, follow the high pointer, $M_H$.

### Multiway Nodes

The node cell must have one pointer for each direction that a node might branch. A possible node design for more than three directions is presented in Fig. 11.5.3. The cell, M, consists of a number of records (or record pointers), each of which is subscripted D. The first is hence $M_{D1}$. There are m of these records, so that the last is labeled $M_{Dm}$.

M

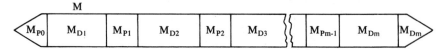

**Fig. 11.5.3.** Data cell for the node of a multiway tree

There is one more pointer than record. The first pointer is $M_{P0}$. Other pointers are labeled according to the data cell that they follow. Thus, $M_{P1}$ is the pointer that follows $M_{D1}$. The last pointer is called $M_{Pm}$.

The records appear in the cell in order of increasing key, so that the record with the lowest number subscript has the smallest key. A pointer falls between two records and points to a branch. This branch contains records all of whose keys fall between those records that bound the pointer.

For example, $M_{P2}$ is between $M_{D2}$ and $M_{D3}$. All the keys in the branch that $M_{P2}$ points to fall between $M_{D2K}$ and $M_{D3K}$.

It should be clear that this is just one extension of the ternary tree just examined. When we arrive at a node, we examine the keys found there. If we find the record, good. Otherwise we find a pointer which lies between two record keys which fall on either side of the key for which we seek.

## 11.6 SEMIDENSE LINKED LIST DIRECTORY

### User Needs

This list structure was specifically designed to fulfill several important user needs and should find important use in the future if an effective system is built around it. ISAM was originally designed to meet the user's requirements to process his file differently at different times. This file structure is an improvement and has been adopted by IBM for the Access Method it calls VSAM, which recognizes that the user at different times wishes to

- process sequentially by key;
- retrieve randomly by key;
- append and delete quickly;

- handle variable length records;
- hold records that grow or contract during use.

This file organization combines several principles that we have already examined:

- Neighborhood locatability enables us to find quickly a subset of records.
- The linked list structure enables us to maintain an ordered file in an unordered list.
- Local overflow provides space for adding new records.
- Further overflow permits expansion of the directory as the list proper expands.

### Single-Level Linked List Directory

This structure is presented pictorially in Fig. 11.6.1. It is very similar to the single-level directory file described in Section 8.3. Thus we find a directory consisting of entries, one for each sublist. The list proper of records is broken into sublists corresponding to the directory entries. The main differences are as follows:

- Each sublist is a fixed area containing a variable number of records.
- The sublist is semidense and space is provided for new records.
- Sublists are linked together.
- The directory is also fixed in size but is semidense and so may contain a variable number of entries.
- The directory permits addition of new directory entries.
- Directories are linked together by pointers.

Since sublists are linked, they need not be contiguous, except internally.

### Searching and Posting

Again we refer to Fig. 11.6.1 to examine how the file is used in various ways.

*sequential*　　Posting requires that we update records sequentially or ·at
*reference*　　least that we examine them sequentially. The list is ordered,
　　　　　　so necessarily within a sublist the records are arranged in key order and a simple sequential search can reveal if the record is in that sublist. If not, the pointer is obtained from the end of the list to the next sublist. It should be clear that garbage must be easily identified or skippable to prevent incorrect interpretation.

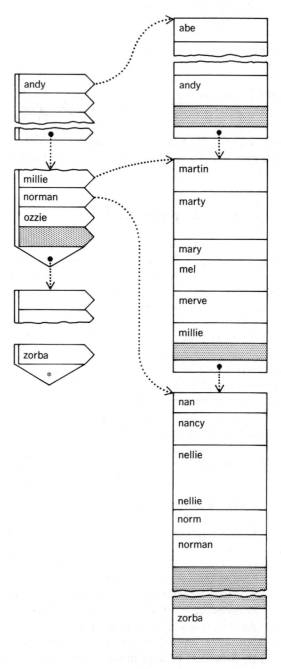

**Fig. 11.6.1.** A semidense single level linked list directory file

334

*random*    To find a particular record such as nat, search the directory.
*entry*    Follow through the entries; if the proper one isn't found,
go on to the next set of entries by following the link. A
search for nat finds an entry for norman that points to the sublist headed by
nan, containing nat.

If we had been looking for ned, we would determine that he is absent when
we reach nellie in nan's sublist.

*variable*    It is easy to see that the records and the cells that contain
*length*    them can be of variable length. During posting we can
*records*    rewrite a record in place, even when the record changes size.

Thus, after referring to nat, we may wish to change or add
fields to him. We can rewrite a larger record in place if we first push down
nellie and norman to provide the additional space. This technique will become
clearer when we examine appending.

### Appending Using Local Overflow

It is easy to append a new record especially if there is room
in local overflow, which appears at the end of the semidense sublist. The
action is illustrated in Fig. 11.6.2. Here it is desired to append a record with
key martin. First search the directory to find that, if martin is present, he is in
a sublist ending with millie. Now go to this sublist to verify that martin is
absent. Sure enough, he is missing, and we note that he should be placed
between marin and marty.

Next check local overflow and find out that there is more space there than
required for the new record. The action performed is the same as used in
maintaining an ordered semidense list for appending. That is, all the records
below marin are moved downward in the sublist by the length of martin, with
millie being moved first: marty, mary, mel, merv, and millie. They are moved
to make room for martin, who is then inserted as the figure shows.

### Appending with a New Sublist

Figure 11.6.3 shows what happens when the record to be
appended is larger than the area left in local overflow. Here we wish to add
nigel. A directory search reveals that nigel belongs in the sublist ending with
norman. Verify that nigel is truly absent and that he belongs between nellie
and norman. There is not enough space for nigel in the overflow area, however.

What to do? First obtain another sublist. Sublists are *linked* together and
it is easy to tack on a new one. Now the records nan, nancy, nat, nellie, nigel,

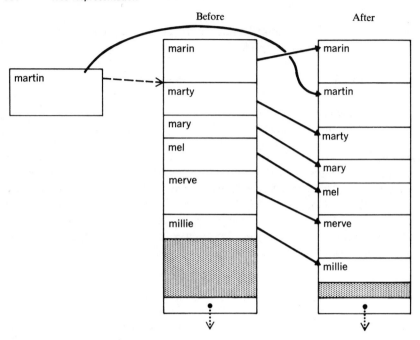

**Fig. 11.6.2.** Adding the new record martin to a semidense linked list directory file, using local overflow

and norman are to be separated into two sublists. We try to do this equitably. If we don't, and we simply move norman to the new sublist leaving the old one still crammed full, then it will still be a problem to append to that old sublist.

Figure 11.6.3 shows that nan, nancy, and nat remain in the old sublist, while nellie, nigel, and norman are put in the new sublist.

**relinking**      The old sublist pointed to the next sublist, beginning with olga. Now we want to insert the new sublist in between. The old sublist now ends in nat and is pointed to the sublist ending in norman; the new sublist ending in norman is pointed to the next old sublist beginning with olga, as shown in the figure.

**directory**      We have created a new sublist for which there is no directory
**entry**          entry and, correspondingly, the directory entry for the altered old sublist is no longer correct. It should record nat as the terminal key.

Revising the directory is illustrated in Fig. 11.6.4. The directory sublist that we use to add nigel contained entries for millie, norman, and ozzie. We

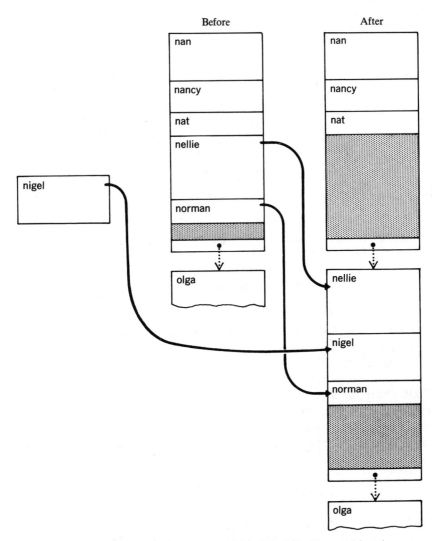

**Fig. 11.6.3.** Adding nigel to the semidense linked list file, requiring the addition and linking of a new sublist

need another entry. Since there is room in directory sublist local overflow, ozzie and norman can be moved down to make room for a new entry for nat. It is a toss-up whether you want to consider nat or norman the *new* entry. Norman is now associated with the new sublist; nat is associated with the old sublist but reflects the key of the terminal record stored there. Whichever way you slice it, the action required is shown in Fig. 11.6.4.

Before                                        After

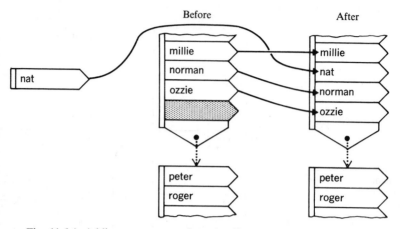

**Fig. 11.6.4.** Adding a new entry into the directory sublist for the new sublist created in Fig. 11.6.3

### New Directory Sublist

We have seen that the addition of a record causes reshuffling of a sublist, which may lead to the creation of a new sublist and hence a new entry in the directory sublist. The overflow area in the directory sublist provides for this new directory entry space, but what happens when we run out of space in the directory sublist?

Of course, another directory sublist is created and we simply divide the entries from the old directory sublist into two parts distributed equally between them. This is illustrated in Fig. 11.6.5, where the new entry for myrtle cannot be placed in the current directory sublist, so a new directory sublist is commandeered.

### Deletion

When a record is deleted from a sublist, order is maintained in the sublist and we close up ranks to eliminate the hole that would otherwise result. Records are moved up and as the dense portion occupied by the records contracts, the bottom portion available for overflow expands.

Eventually a sublist may contract to contain a single record. After we delete *that* record, the entire sublist is empty. To keep our search procedure consistent, we should unchain the empty sublist by linking the predecessor sublist to the successor sublist. We return the empty sublist to the sublist pool.

A deleted sublist requires that the corresponding directory entry is deleted. Again, close up ranks, returning space to the overflow area of the directory sublist.

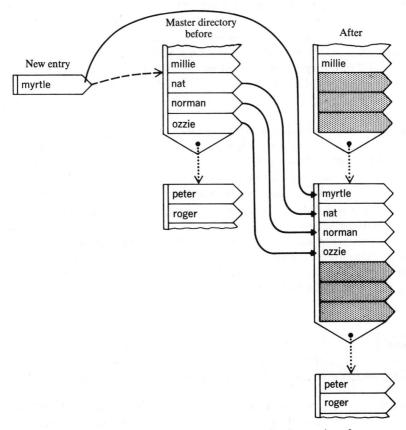

**Fig. 11.6.5.** The addition of a new record may require the creation of a new sublist and also a new directory sublist if there is no room in the proper directory sublist.

Excessive deletion may cause the removal of a directory sublist. Since these are fastened together with links, this too is an easy matter.

## 11.7 MULTILEVEL SEMIDENSE LINKED LIST DIRECTORY TREE

### Multilevel

We have seen in the previous section how semidense directory sublists were used to enter semidense sublists, where both main and directory sublists are linked together. Also both the main sublists and the directory sublists are fixed size. As the number of records in the list grows, so does the number of sublists and, as we would expect, the number of directory sublists increases too. We wish to keep both kinds of sublists at a convenient size. The optimization demonstrated in Section 8.4—making the sublist and directory size dependent on the number of records—is not applicable because optimum sublist size may depend on the media parameters. Still, if the number of records is going to grow, so is the size of the directory.

***example***    Suppose for large records that the best size for a disk sublist is one that accommodates ten records. A list of 10,000 records has 1000 sublists and as many directory entries. With 10 entries for each directory sublist, we end up with 100 directory sublists. (The number of entries per directory sublist may be different and since the entries are small, we may accommodate many more directory entries in its sublist.)

The number of looks depends on the number of entries in the directory, not the number of entries in a directory sublist. Thus, regardless how many entries there are in a directory sublist, on the average we shall have to go half way through the directory, look at 500 entries, to get to the right sublist. The total number of looks is given thus:

$$\hat{L}_D = 1000/2 = 500; \qquad \hat{L}_S = 5; \qquad \hat{L} = 505 \qquad (11.7.1)$$

***remedy***    To remedy this situation, simply place a master directory on top of the original directory. Then it is a simple matter to use the master directory to determine which subdirectory sublist to examine.

***example***    For 10,000 records and 1000 sublists, let us suppose that we have 100 directory sublists. This produces ten master directory sublists and, after doing a little arithmetic, the total number of looks is reduced from 505 to 60, determined thus:

$$\hat{L}_M = 50; \qquad \hat{L}_D = 5; \qquad \hat{L}_S = 5; \qquad \hat{L} = 60 \qquad (11.7.2)$$

### Structure and Search

The two-level semidense linked list directory tree is very much like the two-level directory file. The important difference is that the sublist, subdirectory, and directory are all semidense:

- There is probably room in each sublist for additional records.
- There is probably room in the subdirectory sublist for additional entries should another sublist be required.
- There is probably room in the directory sublist for new entries should a new subdirectory sublist be required.

Figure 11.7.1 illustrates how the two-level directory file of Fig. 8.5.1 is converted into a two-level semidense linked list directory tree.

It should be clear that search proceeds as with the two-level directory except that when we reach the end of a directory sublist, we use the link to find the

**Fig. 11.7.1.** A multilevel semidense linked list directory tree

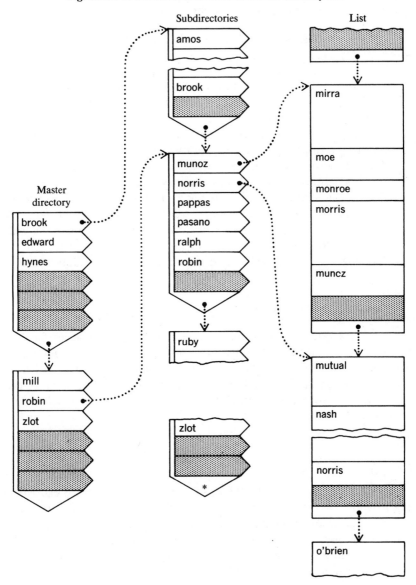

next sublist. This is true of subdirectory search also except that we should not have to use the link since the directory localizes which subdirectory to enter. Finally, the subdirectory takes us to the proper subsublist where either the record is found or we verify that it is absent.

### Appending and Deleting

It should be clear how appending and deleting are done. We shall only examine appending cursorily. To append a new record:

1. Find the subsublist in which it belongs.
2. Rearrange the records to keep them in order, using overflow for the additional space required.
3. If there is not enough space in overflow, obtain another subsublist, split up the records, and link the subsublists together.
4. Another subsublist means an additional subdirectory entry—go back to the subdirectory and fabricate the new entry there.
5. Should the addition of the new entry require creation of a new sub-directory, obtain a new sublist and split the original subdirectory. This entails the addition of a new entry in the directory—create this entry and properly enter it.
6. The directory may grow so large as to require another sublist for it.

### Higher Levels

As with higher level directory files, higher levels facilitate the use of the semidense linked list directory tree. It may be made an automatic feature in the construction of the tree. Thus whenever the directory contains multiple sulists, a higher level directory is created to point to each of the directories on the lower level. Hence we can pile directories upon directories without limit.

### Numbers

It is clear that the length of the record generally exceeds the length of the subdirectory entry. It is useful to gather records together in blocks to suit a DIRECT ACCESS STORAGE DEVICE. Hence it is often convenient to make the block size equivalent to the sublist size. Therefore there will be many more entries in a subdirectory sublist than sublists of the list proper.

This inequality may not be totally satisfactory. It is always up to the file designer to determine the best tradeoffs between the time required for medium access and the time required for sublist search:

- DISK reposition and access consumes milliseconds.
- COMPUTER sublist processing takes microseconds.

We must also consider that while the disk is positioned, the COMPUTER may be used by other programs and the time will not be charged to us.

## PROBLEMS

11.1   Use Table 8.P.1 which is L, and make this list into a binary tree $L'$. The cell $L'_i$ contains two pointers, $L'_{iL}$ and $L'_{iH}$, to the low and high nodes, respectively, on the next level. For convenience, put these in decimal as the last two fields in $L_i$ so that we have

$$[L'_{iL}] = [L_i] + 90$$
$$[L'_{iH}] = [L_i] + 95$$

11.2   Make a chart showing all $L'$.

11.3   For the file in Problem 11.2, make a binary tree total directory (BTTD) by creating the tree directory T. An entry $T_i$ consists of
- a key field $T_{iL}$ of five bytes,
- a data pointer $T_{iP}$ of five bytes,
- a low pointer $T_{iL}$ of one byte,
- a high pointer $T_{iH}$ of one byte.

The tree pointers give in binary the ordinal number of the next entry $T_j$ in T, so that $[T_j]$ can be calculated from $T_{iL}$ or $T_{iH}$.
   (a) Show T where $[T] = 30,000$ and where $T_{iL}$ and $T_{iH}$ are shown in decimal though actually in binary.
   (b) Indicate how to find $[T_j]$ given $T_{iL}$.

11.4   Use T to find the names in Table 11.P.1 in L.

**Table 11.P.1.** Names for Look Up

| | |
|---|---|
| Goldin | Falco |
| Goldstein | Farella |
| Goldman | Fiore |
| DeFalco | Flores |
| Marsden | Marlboro |

11.5 As long as **L** is not subject to change, the position of $T_i$ in **T** is equivalent to the position of $L_i$ in **L**. Hence $T_{iP}$ can be omitted. Consider the structure of **T′** where $T'_{iP}$ is absent from $T'_i$ and *len* $T_i = 7$.
(a) Sketch **T′**.
(b) Provide the formula to find $[T'_j]$ from $T'_{iL}$ or $T'_{iH}$.
(c) Provide the formula to find $L_i$ from $T'_i$.

11.6 Use **L** as amended in Problem 11.1 and find the records with keys in Table 11.P.1 using **T′**.

11.7 Compact **L** of Table 8.P.1 by rewriting it as **G** where $G_i$ consists of
• a variable length name (key) field terminated by $;
• a 10-character data field.
Create **U**, a BTTD for **G** with an entry composed thus:
1. five-byte truncated key;
2. five-byte data pointer;
3. one-byte low ordinal pointer field;
4. one-byte high ordinal pointer field.
Put **G** at 60,000 and **U** at 80,000.

11.8 Look up the names in Table 11.P.1 using **G** and **U**.

11.9 Prepare **T** for balancing by creating **T″** where $[T''] = 25,000$.
$T''_{iK}$ and $T''_{iP}$ are key and pointer fields of five bytes.
$T''_{iB}$ is a balance field of one byte: $00_H$ for balanced, $80_H$ for left unbalanced, $01_H$ for right unbalanced (the subscript H denotes hexadecimal).
$T''_{iR}$ is a reverse pointer byte giving ordinal number.
$T''_{iL}$ and $T''_{iH}$ are one-byte ordinal number low and high pointers.
$L'T''_i = 14$ bytes.

11.10 Add the records in Table 11.P.1 to **L** (Problem 11.1) in the order shown in cells of 100 bytes starting at 54,600. As each is entered, put a new entry, $T''_X$, into **T″** and fill $T_X$. (Determine $T_n$, the node where $T_X$ is attached, and reset.) Then check **T″** for balance. If necessary, rebalance **T″** by resetting any of $T''_{iB}$, $T''_{iR}$, $T''_{iL}$, and $T''_{iB}$ for any $T'_i$.

11.11 Prepare **U′** at 85,000 for **G** but where **U′** has the same format as $T''_i$.

11.12 Add the variable length records in Table 11.P.2, this time to **G** as described in Problem 11.7. Alter **U′** of Problem 11.11 for each incoming record as described in Problem 11.10.

11.13 Delete records from **L** but only by altering **T″**. Use **T″** from Problem 11.10 and rebalance the tree as each record is deleted. The delete list is in Table 11.P.3.

11.14 As for Problem 11.13, delete the records listed in Table 11.P.3 from **G** by altering **U′**.

11.15 For Table 8.P.1, construct a ternary tree, **Y**. Use the entry format in Table 11.P.4. Place the tree **Y** starting at 90,000.

**Table 11.P.2.** Additions to L and G

Globe Sound Inc.
Foti Arthur
Frisch Irwin
Glinka S
Goodhue N T
Four Guys Orchestra
Fox M L & Son
Gloria's Doll Hospital
Friedman Nat
Gelish Paul
Global Planning Corp.
Glenwood-Penn Inc.
Gristede Bros.
Glack Sanford
Gold B D
Golden Age Nutrition
Glugs Gary F

**Table 11.P.3.** Deleted List

| | | |
|---|---|---|
| Iuliucci | Gold | Giamo |
| Macek | Ding | Giglio |
| Hall | Dunne | Gilmartin |
| Glass | | |

**Table 11.P.4.** Ternary Tree Entry

| Field | Purpose | Length |
|---|---|---|
| $Y_{KL}$ | Key, low | 5 |
| $Y_{DL}$ | Data pointer, low | 5 |
| $Y_{KH}$ | Key, high | 5 |
| $Y_{DH}$ | Data pointer, high | 5 |
| $Y_L$ | Tree pointer, low, ordinal number of cell | 2 |
| $Y_M$ | Tree pointer, medium, ordinal number of cell | 2 |
| $Y_H$ | Tree pointer, high, ordinal number of cell | 2 |

11.16 Use the ternary tree, $Y$, to look up the records in Table 11.P.1.

11.17 The records with keys in Table 8.P.1 are to be placed on a DASD as in Fig. 11.P.1 starting at track 100. Each track holds two blocks; each block holds eight records. Only six records are placed in each block now to provide local overflow; nineteen tracks have been so formatted. A directory J is set up at location 80,000 in MEMORY. It is also semi-dense with space for eight entries per block, only six of which are

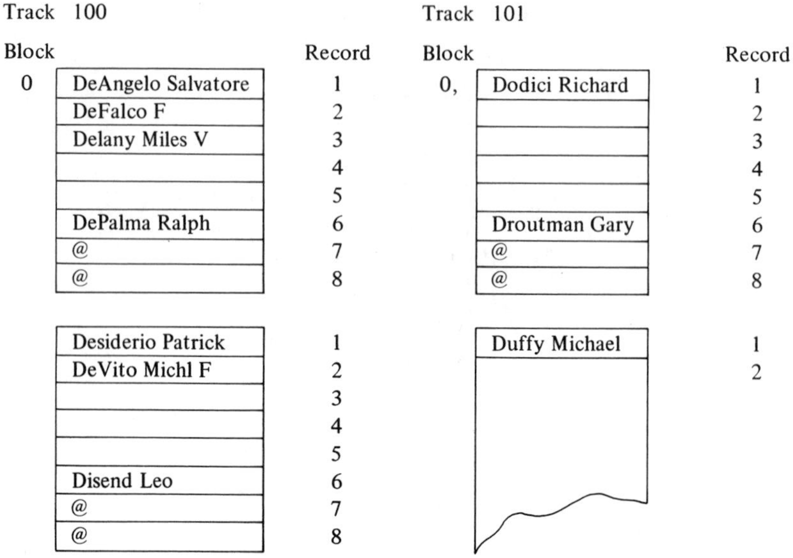

Track  100                                    Track  101

| Block | | Record | | Block | | Record |
|---|---|---|---|---|---|---|
| 0 | DeAngelo Salvatore | 1 | | 0, | Dodici Richard | 1 |
| | DeFalco F | 2 | | | | 2 |
| | Delany Miles V | 3 | | | | 3 |
| | | 4 | | | | 4 |
| | | 5 | | | | 5 |
| | DePalma Ralph | 6 | | | Droutman Gary | 6 |
| | @ | 7 | | | @ | 7 |
| | @ | 8 | | | @ | 8 |

| | | Record | | | | Record |
|---|---|---|---|---|---|---|
| | Desiderio Patrick | 1 | | | Duffy Michael | 1 |
| | DeVito Michl F | 2 | | | | 2 |
| | | 3 | | | | |
| | | 4 | | | | |
| | | 5 | | | | |
| | Disend Leo | 6 | | | | |
| | @ | 7 | | | | |
| | @ | 8 | | | | |

**Fig. 11.P.1.**

occupied. Each entry consists of an eight-character truncated key and a four-character DASD block pointer. There is a link pointer of five characters to the next directory block. Display the directory.

11.18 Add the records for which the keys are in Table 11.P.2. Use empty DASD and directory blocks where necessary. Display those DASD and directory blocks which have changed or which are new. (For J and L)

11.19 Delete the records in Table 11.P.3. (For J and L)

# CHAPTER 12

# Complex Files

## 12.1 INTRODUCTION

### Definition

Many file organizations have been examined in early chapters. For instance,

- the sequential file;
- the linked list;
- the directory system;
- the binary tree.

A **complex file** is one where several of the record relations prevail simultaneously. Complexity arises from the need to perform more than one specific action upon the file in the least amount of time and where, further, the actions to be performed are predictable and known ahead of time.

*example*    An example of a complex file is a blocked linked list. We have described an advanced modification of this as the semidense linked list directory. The **blocked linked list** is a combination of a serial list and a linked list, as shown in Fig. 12.1.1. Here the solid line indicates the physical proximity of one record to the next; the dotted line indicates a pointer from one record to the next. For the blocked linked list, a number

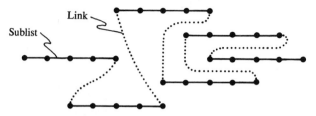

**Fig. 12.1.1.** The blocked linked list file

of records are physically associated in a sublist (or block). At the end of the sublist there is a pointer to the next sublist where the file continues.

Below two complex file structures are introduced. These are pursued in more detail in later sections.

### Multilist File

The multilist file is an important complex file. We have seen as required for the directory system how a file is split into several subfiles, each of which contains several whole records; each subfile occupies a sublist of the main list. This **horizontal partitioning** is illustrated in Fig. 12.1.2. To

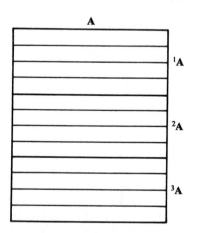

**Fig. 12.1.2.** Horizontal partitioning

see the origin of this term, string out the records $r_1, r_2, \ldots, r_n$ of the file into a long vertical string. Then make a series of horizontal cuts, each between two records, dividing the vertical string of records in several pieces. These cuts define a number of subfiles. Each subfile contains a number of *complete* records. $_1A$ contains records $r_1, r_2, r_3$ and $r_4$; $_2A$ contains $r_5, r_6, r_7$ and $r_8$; and so forth.

The **multilist file** is created by **vertical partitioning**, as shown in Fig. 12.1.3. As before, string out the records $r_1, r_2, \ldots r_n$ of the file vertically. Now cut the file *vertically* so that each of the resulting sets contains *one* part of *every* record. Thus each subfile has as many members as the original file. In the figure, $_1B$ contains the first few fields of all the records $r_1, r_2, \ldots, r_n$; $_2B$ contains the next few fields of all the records $r_1, r_2, \ldots, r_n$; etc.

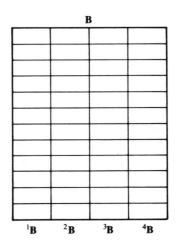

**B**

**Fig. 12.1.3.** Vertical partitioning    $^1B$    $^2B$    $^3B$    $^4B$

Sometimes the strings thus created are called header and trailer files. A **header file** consists of subrecords, each of which contains a record identifier and at least a pointer to its continuation subrecord in the **trailer file**.

### Multilink File

Another important complex file is the **multilink file**, a multiple extension of the linked list concept. It is the application of the linking principle a multiple number of times to a single file. Now each record, instead of having just a *single* link field that relates it to other records in the file, has several other link fields. Each link field stands for a different attribute or relation that might exist among the records. Thus one link field may serve to relate records together because of the individual's age. Another link field may relate records together by occupation.

*others*     There are other complex list formats. We shall not examine all possibilities; an encyclopedia would be required for this. Instead, we examine the ramifications of the multilist file and multilink file in the sections that follow.

## 12.2 MULTILIST FILE

### Need

Often some of the information that describes the individual is more important to the user than other information. This **active information** may be used for reference or may require changing during a frequent activity, such as posting. In any case, when using the active information, it is inefficient and time-consuming to pass through or review the **inactive** (for part of the application) information.

The **header (sub) file** (or, sometimes, the **header list**) is a collection of subrecords, one from each record in the file. Each subrecord in the header file contains an identifier or key field and active information that is subject to reference or change.

Here subrecord takes on a different menaing. A **subrecord** is a collection of fields as before. But the fields are not associated because they apply to a multiple attribute; they are associated because they describe a type of attribute—here active or inactive.

The **trailer (sub) file** provides the inactive (or less active) information about the individual. There may be a number of such trailer subfiles, each consisting of subrecords with descriptions applying to one or more individuals in the universe and hence to one (or more) subrecord(s) in the header subfile.

I have drawn your attention to the fact that a user *file* has been broken down into *subfiles*. Also, since the partition is *vertical* rather than horizontal, *records* are divided into *subrecords*. Hence *subfiles* now are sets of *subrecords*. Common usage drops the "sub" to speak of a header and a trailer file and their associated records. I shall now revert to common usage but caution the reader to be aware of the proper meanings.

Supposedly, each record in a trailer file, **tr**, is associated with at least *one* record in the header and perhaps more. We can describe the trailer file according to how many header records they apply. When each trailer record corresponds to exactly one header record in the header file, **hd**, and vice versa, as in Fig. 12.1.3, this is a one-to-one trailer file and

$$num \textbf{ hd} = num \textbf{ tr} \tag{12.2.1}$$

*many-to-one*   A trailer file may have one subrecord that applies to several different records in the file or to none in some instances. In fact, one advantage to the multilist file may be that information duplicated in several records can be abstracted from those records and placed in a single trailer record in a many-to-one trailer file where it now applies to all the headers. The vendor file in the inventory application is a good example.

For the header file **hd** and the trailer, the many-to-one relation is described thus:

$$num \ \mathbf{hd} > num \ \mathbf{tr} \qquad (12.2.2)$$

***one-to-many***   There may be several trailer records associated with a single header record. Thus, for a checking account, the header record may contain key and summary information; there is one trailer record for each check issued or deposit made. Therefore although each header may have *several* trailers, any trailer is associated with only *one* header. Then we have

$$num \ \mathbf{hd} < num \ \mathbf{tr} \qquad (12.2.3)$$

***many-to-many***   Again, several trailers belong to one header, however, one trailer is *not* now bound to a *single* header. A trailer may go with several headers. Consider a parts explosion. One assembly may have many parts and subassemblies. But the same part or subassemblies may be used with many assemblies!

### Makeup

The makeup of the multilist file in more generalized terms is shown in Fig. 12.2.1. The file is spread over several sublists. The sublists comprise the list—the space that is occupied by the file. There is no need for the list to be contiguous. In fact, each sublist might be on a different volume.

The header list by definition contains one identifier subrecord for each

**Fig. 12.2.1.** The multilist file

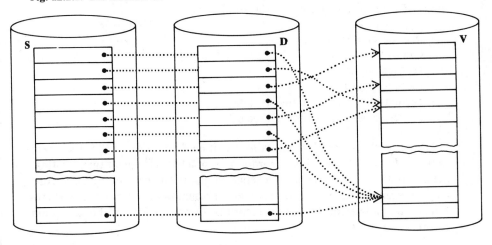

record in the file. The trailer list may be one-to-one, many-to-one, etc. In any case, there is a pointer from the header record to each applicable trailer record in its own trailer list. Clearly, for the many-to-one case, there are pointers from several header records (or major trailer records) to a terminal trailer record.

### Inventory Example—Subfiles

Let us consider an example as the best means for presenting the details of one form of the multilist file. For the inventory application, the header record identifies each stock item along with quantities in stock and so forth. Hence, the header list is the stock list (which contains the stock file); designate this stock file as **s**. Each item in inventory has a description record that applies to it in a one-to-one relationship. The description file designated as **d** contains a unique description record for each header record and we have

$$num \ \mathbf{s} = num \ \mathbf{d} \tag{12.2.4}$$

Associated with each item is a vendor from whom that item is purchased. A vendor may supply a number of different items. The vendor file is designated **v** and we have

$$num \ \mathbf{s} > num \ \mathbf{v} \tag{12.2.5}$$

There is one stock record, s, in the stock, **s**, for each item in inventory. There are a number of fields contained in each such record for the purpose of keeping an accurate account of the current inventory and for determining whether to reorder and how many items to reorder when necessary. At least these fields should be present in the stock record:

- item identification;
- quantity on hand;
- reorder point;
- reorder quantity;
- current reorder action and its date;
- address of the descriptor subrecord.

The description record, d, contains a written description of the stock item to be used in reordering. Since reordering is much less frequent than posting, the description is maintained in a separate place. The description subrecord may contain other important reorder information not found elsewhere. Finally, the pointer to the vendor record is found in a field of the desription record.

The vendor list, **v**, contains one record, v, for each vendor. This record contains the name, address, and other information such as might appear at the top of the purchase order.

*organization*    The inventory list thus contains three lists: the lists called **S** and **D** each contain one subrecord for each item. Hence these two lists have a one-to-one relation. A vendor may supply a large number of items to us, however. Vendor information need appear only in the vendor file. Hence, the vendor list, **V**, bears a one-to-many relation to other lists.

### Posting

Consider the stock file, **s**, and a transaction file, **t**. There is one record on **t** for each entry or withdrawal of items from inventory. The purpose of posting is to record this entry or withdrawal upon the stock record. The first action of the posting activity is to relate each transaction record, t, to some stock record, s. We have found that this is best done where both files are ordered. The action produces a file of active stock records **a**, indicated thus:

$$\mathbf{t} \; esearch \; \mathbf{s} \longrightarrow \mathbf{a} \qquad (12.2.6)$$

Of course, a file of reject records, **r**, may be created when the transaction record cannot be matched in the stock file. Then there are inactive stock records, **p**. We have

$$\mathbf{t} \equiv \mathbf{a} \cup \mathbf{r}; \quad \mathbf{a} \cap \mathbf{r} \equiv \Lambda; \quad \mathbf{s} = \mathbf{a} \cup \mathbf{p}; \quad \mathbf{a} \cap \mathbf{p} = \Lambda \quad (12.2.7)$$

where $\equiv$ means *corresponds to*.

Next, the transaction is posted into the active record, which is indicated using the update operator, *up*, thus:

$$\mathbf{t} \; up \; \mathbf{a} \longrightarrow \mathbf{a}' \qquad (12.2.8)$$

There are two kinds of activities that are recorded:

- *withdrawal* (from inventory), where the clerk has removed items and recorded his activity with a transaction record, t;
- *entry*, where new items shipped by the vendor are placed onto the shelf by the clerk who also creates a transaction record, t, for this action.

Posting consists of subtraction for the first case and addition for the second case, of the transaction amount from the on-hand field in the record s.

***reorder***    After an active stock record is posted, it is necessary to check the on-hand field against the **reorder point**, a more or less permanent field in s. If the amount on hand has gone below this point, we should place an order to replenish our stock. The amount to be ordered, called the **reorder quantity**, is also found in s. Before issuing a new order, however, check the current order information in s. If a previous order was placed during earlier posting activity, there is no need to issue a duplicate.

If there is no new order in the works, a new order skeleton record, n, is created and placed in the new order skeleton file, **n**.

### New Order Processing

The new order record, n′, prepared after the posting activity, consists of two parts:

- the reorder quantity information and the item identification obtained from the posting activity and found in **n**;
- the description obtained from the description file **d**.

Search for and attachment of the description record d can be done during posting or on a separate run. Let us call the file of active descriptions **d′**. It is obtained thus:

$$\mathbf{n} \; esearch \; \mathbf{d} \rightarrow \mathbf{d'} \tag{12.2.9}$$

Each new order skeleton, n, is now combined with an active description, d′, to form a new order record, n′:

$$n' = n \cup d'; \quad \mathbf{n'} = \{n'\} \tag{12.2.10}$$

where $\cup$ means *concentrate or paste together*.

We now have an order specification and a complete item description, **n′**, for each item to be reordered. We associate a group of these orders and descriptions with a particular vendor who will supply the items to produce a purchase order (po). To perform this association efficiently, sort the new descriptor file **n′** by vendor identification. We shall then have one or more records, each of which represents an item in a group to be supplied by a single vendor. Take the file, **n′**, which is in order by item number. Designate a *different* order key field, namely vendor identification. Set out the file on this key to create a new file, **n″**. It is a special kind of ordered file since there may be several records with the same key! For a given (vendor) key, however, the order of the records in the group is unimportant.

*collating*     Now the sorted new order file, **n″**, is collated with the vendor file to provide the active vendor file, **v′**, which might be indicated thus:

$$\mathbf{n}'' \; esearch \; \mathbf{v} \longrightarrow \mathbf{v}' \tag{12.2.11}$$

An active vendor record, **v′**, is combined with one or more new description records **n″** to create a purchase order record po from which we shall print out the purchase orders.

Since **n″** is a special kind of ordered file, possibly several records with the same key, the active vendor file **v′** contains *one* record for *one or more* records in **n″**. Now we combine **n′** and **v′** to create **po**, the purchase order file. Then **po** can be structured in two ways:

- as a multilist file;
- as a mixed or compound sequential file (Section 12.3).

Regardless, the creation of **po** can be symbolized thus:

$$\text{po} = \mathbf{v}' \cup \mathbf{n}_1'' \cup \mathbf{n}_2'' \cup \cdots \cup \mathbf{n}_m''; \quad \mathbf{po} = \{\text{po}\} \tag{12.2.12}$$

The record po in **po** conveys all the data for a single purchase order:

- the vendor identification;
- all vendor data: name, address, etc.;
- item requirements by identification number: item description, quantity, price, etc.

The final stage of the inventory posting activity is to print the purchase orders. This probably requires a special form in the PRINTER with a special carriage tape so that the purchase orders can follow a standard format.

## 12.3 THE COMPOUND FILE

We have encountered the compound file earlier in one or more of its alternate forms. The **compound file** contains two or more types of records and/or several records with the same key. If the compound file is not in key sequence, it is truly a mess.

### Ordered Compound File

When a compound file is in key sequence, it is called **ordered compound file**. Additionally, where different types of records occur, it is customary to require the record types to be in a particular order. Hereafter we shall refer to the *ordered compound file* simply as a **compound file**.

Figure 12.3.1 portrays the compound file. Records of two types are present. Let us call the type that appears first and is represented by a square a **major** (type) **record**. The other record type (circle) is a **minor record** (type).

**Fig. 12.3.1.** The compound file

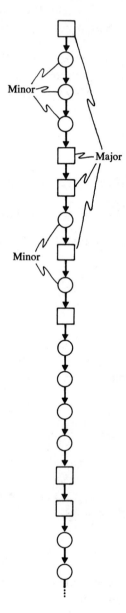

In the figure the arcs connecting the records represent the physical sequence of the records in the file.

### Subrecords

Another way to look at the compound file is to consider it to be a simple file consisting of compound record. Here a compound record consists of multiple subrecords. In its favorite form a **compound record** consists of a **major subrecord** and none, one, or several **minor subrecords**.

Referring again to Fig. 12.3.1, each compound record, under this other conception, begins with a major subrecord (square) followed by none, one, or more minor subrecords (circle). Naturally, the compound record is of variable length even if its subrecords are of fixed length.

### Order

When we conceive of the compound file as a simple file composed of compound records, order is less of a problem. Now there may be several subrecords with the same subkey but there is only one record with that given key. Generally the major subrecord must come up first.

## 12.4 MULTILISTS

### One-to-One Multilist File

The origin of the one-to-one multilist file is from posting where each record consists of an important part which is subject to alteration and the less important part which is only used occasionally in exceptional conditions. We have called the important part the **major subrecord** and the less important part the **minor subrecord**. This is illustrated in Fig. 12.4.1 where the oval represents the record and the left-hand section of the record is called the major subrecord. The vertical line divides the record into the major and minor subrecords.

Let us take the major subrecord of each record and assemble all of them into a single file, as shown in Fig. 12.4.2. This becomes the header file. The remaining portion of the record, the minor subrecord, is placed into another file, the trailer file. Since there is one portion of each record in each file, the multilist is said to be one-to-one.

There is an obvious advantage to this technique. During posting we need only examine the header file. Only when posting causes an exceptional condi-

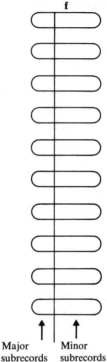

**Fig. 12.4.1.** Splitting records into    Major    Minor
major and minor subrecords    subrecords    subrecords

tion is recourse to the trailer file necessary. During a later activity, the pertinent trailer records can be retrieved and acted upon.

*relation*        Figure 12.4.2 shows the header and trailer files, **h** and **t**,
                both as ordered sequential files. Look up for either files
can be by any of the techniques discussed earlier.

Another alternative is to include in each header record a pointer to the
corresponding record in the trailer file. This may be expressed symbolically
thus:

$$point\ \mathrm{h_i} = [\mathrm{t_i}] \tag{12.4.1}$$

There are no restrictions on the choice of fixed or variable length for
header or trailer records. We often find, however, that header records are
fixed to facilitate access.

### Many-to-One

Again, consider a file for which each record can be divided
into a major and minor subrecord, as in Fig. 12.4.1, and for which we have

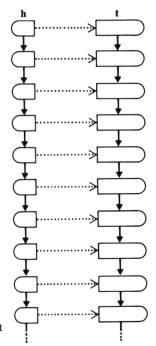

**Fig. 12.4.2.** The one-to-one multilist file

performed a vertical cut, as in Fig. 12.4.2. Upon so doing, it may become apparent that the collection of minor records has many duplicates. If this is so, it is wasteful of space to carry the duplicates, especially since each now has been disassociated from its original major subrecord. Let us remove all the duplicates from the trailer file. If the original trailer file is **t**, let us call the new trailer file without duplicates **t'**. Hence **t'** is designed thus:

$$\mathbf{t'} = \{t_i\} \qquad \text{where } t_i \neq t_j \qquad (\text{all } i \neq j) \qquad (12.4.2)$$

We now have a header and trailer file where the number of records in the trailer file is much reduced. For each record in the trailer file there may be several header records that point to it. The advantage to this is clear—the trailer file is much smaller now. The trailer record cannot be identified by a particular header record, however, since there may be several such header records that claim it. Therefore, a necessary arrangement is for each header record to carry a pointer to its trailer. The pointer can be the physical location of the trailer record or it can be another identifier unique to each trailer record.

The relation of the two files is illustrated in Fig. 12.4.3. To find the proper trailer record, the pointer operator *point* is applied to the header record as in (12.4.1).

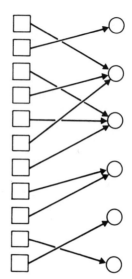

**Fig. 12.4.3.** The many-to-one multi-list file

### One-to-Many

*origin* For each header record we could postulate several trailer records. This structure would occur when the original record is subject to a partition that creates one major subrecord and several minor subrecords. This is done with several vertical cuts. Again, the advantage to be gained by separating the header from the several trailer records belonging to it is to decrease the space occupied by the more often used header records and consequently the time required to scan through the header file.

First postulate several trailer records belonging to a single header record but for any trailer record there is one, and only one, header record associated with it.

The original subrecords and the trailer records that correspond to them can be either fixed or variable in size and may be of either fixed or variable number. Let us discuss the most general case where a variable number of records of variable size are found in the trailer file for each header record. There are at least two ways in which we can structure the trailer file.

*sequential* In Fig. 12.4.4 we see a sequential trailer file. The header
*trailers* record points to the first trailer record by giving its location using a pointer. The remaining trailer records are positioned adjacent to each other. This might be indicated symbolically thus:

$$point\ h_i = [t_{i1}]; \qquad [t_{i2}] = [t_{i1}] + len\ t_{i1} \ldots ; \qquad [t_{ij}] = [t_{ij-1}] + len\ t_{i,j-1}$$

$$(12.4.3)$$

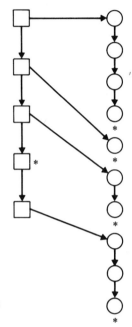

**Fig. 12.4.4.** One-to-many multilist
file with sequential trailers

*linked*          The disadvantage of this sequential trailer file is apparent
*list*            when we decide to add a new trailer record. The entire file
                  must be revised. If the trailer file is kept as a linked list
format, as shown in Fig. 12.4.5, new records can be added at the end of the
file and associated trailer records are simply chained together with pointers.
Thus each trailer record contains a pointer to the next as might be conveyed
thus:

$$point \ h_i = [t_{i1}]; \quad point \ t_{i1} = [t_{i2}]; \quad \ldots \quad (12.4.4)$$

*example*         A checking account application would lend itself to both
                  structures. Here each header record represents an account.
It contains activity information for the month. Individual information about
deposits and checks issued is contained in the trailer file. This file can be
reached with one or more pointers. It is used to produce the statement for
the account holder.

### Many-to-Many

Each header record may be associated with several trailers
and trailers may also be associated with several headers. An example of this
is a parts list. There is a header record corresponding to each assembly,

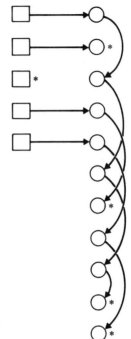

**Fig.  12.4.5.** One-to-many  multilist
file with linked list trailers

subassembly, or part belonging to the mechanism. The same part, however, may be used on several subassemblies. One way to keep such a file is shown in Fig. 12.4.6. Here there is a separate pointer field in the header record for each possible trailer record to be associated with the header. This requires

**Fig. 12.4.6.** Many-to-many multilist
file

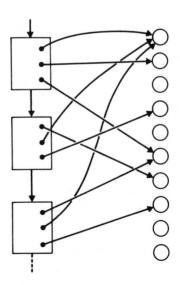

either a large header record or a variable length header. Separate pointer operators, *point*₁, *point*₂, etc. extract each trailer record location from the header thus:

$$point_1\ h_i = [t_{i1}]; \qquad point_2\ h_i = [t_{i2}]; \qquad \cdots \qquad (12.4.5)$$

It is possible to combine the linked list structure where the one-to-many relation holds with a multiple structure, as shown in Fig. 12.4.7.

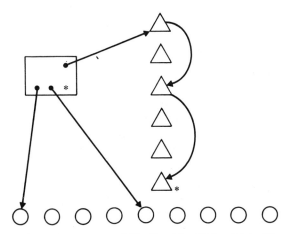

**Fig. 12.4.7.** Many-to-many multilist file with additional linked list trailers

## 12.5 ALTERING AND MAINTAINING MULTILIST FILES

A multilist file, like any other file, can be used in three different ways:

- *reference* where the record is not changed;
- *posting* where only a portion of the record is subject to change;
- *alteration* where new records are created, old ones are deleted, or changes are made in portions not otherwise subject to change.

### Posting

We have defined posting as an activity that alters specific fields in the record. Our file structure was conceived so as to group together in the header record those fields that are subject to change during posting.

Then we can expect posting to affect our multilist file generally in one of two ways:

- Only the header record is altered;
- The header record is altered and trailer records are appended or deleted.

**header only**    The inventory file is an example where posting only alters the header record. The number of items deleted from or added to stock is used to modify the existing header record. This is then examined to determine if the quantity on hand has reached the reorder point. Then a different action will be taken. This action usually consists solely of reference to the trailer file.

*Alteration* of the trailer file is only done during maintenance—when a new vendor is chosen, for instance.

**trailer**    In the checking account example, the header file records
**altered**    the current status of the account. This facilitates random entry to service the account holders' inquiries, for instance. During the (daily) posting period the header record is updated. The trailer record holds information about each transaction for this account. Therefore, during updating, new records are normally appended to the trailer file. The way this is done depends on the organization of the trailer file.

In Fig. 12.5.1 we see a new trailer record being added to the trailer file for an active account where the trailer file is kept sequentially, records being grouped by account number. As you see, trailer records that follow the active account are be advanced to make room for the new record. This technique is unwieldly when trailer files are kept on DASD's. It is a good tech-

**Fig. 12.5.1.** Altering header and sequential trailers

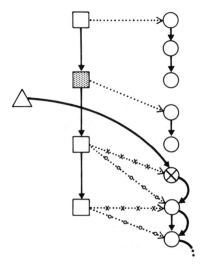

nique, however, when using magnetic tape files. Then during posting the trailer file is updated by simultaneously copying it from one tape to another, adding the new records in so doing.

Figure 12.5.2 shows updating by adding a new trailer record when the trailer file is kept as a linked list. Here the new trailer record is placed at the end of the trailer file (or in intermediate overflow, as we shall examine later). This does not require reorganization of the trailer file and is hence more appropriate for use with DIRECT ACCESS STORAGE DEVICES.

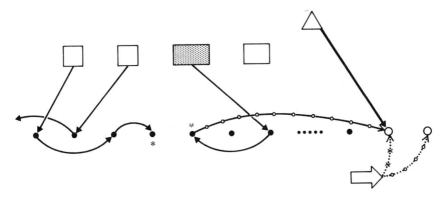

**Fig. 12.5.2.** Altering header and linked list trailers

### Appending

The action of appending depends on the file structure and the quantitative relation between the header and trailer file.

*one-to-one*     When there is one and only one header for each trailer, then appending a header entails appending a trailer for it. The header is placed in its file and positioned to keep order if any. The trailer is put in its file, generally at the end. The header's pointer to its trailer can thus be established. An example is the payroll header and trailer.

*many-to-one*     When a new header record is added to this kind of file, it is simply pointed at the corresponding trailer. Thus in our inventory example where the vendor file is concerned, a new header will be pointed to the existing pertinent vendor record. If the vendor is also new, this entails appending a vendor record too. The vendor record may need to be positioned if the vendor file is ordered by, say, vendor number.

*one-to-many*     When a header corresponds to several trailer records, it is always necessary to add the head in its proper order. It may also be necessary to alter the trailer file by appending trailer records to it.

A new checking account record is added to the header file. Probably no trailers come with it if no activity (checks or deposits) is recorded for it.

***many-to-many***    In this case we have both one-to-many and many-to-one relations existing simultaneously. The requirements of each must be fulfilled. To add a new assembly (header) to a manufactured item, all subassemblies and basic parts (trailers) must be accounted for. New subassembly and part trailer records are needed if absent.

### Deleting

The death of a header may or may not take any trailer records with it. It will for one-to-one and one-to-many; it may not for many-to-one or many-to-many. A dead header record may be removed from the header file to decrease space and the consequent time for processing. Since the trailer file is less used, there are three ways to handle a dead trailer:

- Mark it and "expunge it."
- Mark it as "deleted" but keep it in the file.
- Ignore it—if all headers pointing to it are dead, it won't be referenced.

### Intermediate Overflow

Intermediate overflow is useful when the trailer file is subject to frequent appending, especially for a one-to-many relation as exemplified in Fig. 12.5.2. If the header and trailer files are kept on DASD's, then adding new records into the trailer file would normally place them at the end of the file in a different cylinder from the other records belonging to a given header. Instead, trailers associated with a group of headers may be kept in the same cylinder with intermediate overflow provided within that cylinder. Then when a record grows because of additional subrecords newly associated with it, the new subrecords can be placed in the same cylinder. They can then be chained together without materially increasing access time.

### Reorganization

Depending on the file structure chosen, it is more or less effective to perform a reorganization of the file from time to time. Reorganization is helpful when much appending and deleting takes place and holes and disorganization tend to occur in the trailer file.

In some cases, reorganization takes place automatically. Consider the checking account application using magnetic tape. The header record represents a summary of the account while the trailers describe each individual activity on the account. The information in these files is maintained during the month and is updated daily or as promptly as possible to describe the status of the account so that overdrafts can be prevented. (This is essential in a savings account application.)

At the end of the month the account holder is sent a statement that includes the summary information in his header record and a printout of all the trailer records. Thereafter the trailer file as it now exists is no longer required. It would be cleared out in preparation for a new trailer file starting with the first business day of the month. It would be best copied onto a backup tape. The header record, which may contain such information as the number of checks and the number of deposits made during the month, is reset but not eliminated.

For the inventory example one trailer file contains vendor records. As a particular vendor is phased out, less and less header records point to that vendor. Eventually there will be no headers pointing to the trailer for this vendor now in disfavor, which will be maintained until reorganization takes place to clear him out.

The frequency and amount of reorganization depends on the organization of the file, the frequency of use, and the percentage of turnover during each posting and maintenance period. There is no simple formula to describe it.

## 12.6 MULTILINK LISTS

### Subfile Partition by Attribute Value

We have examined the use of a linked list to provide an ordered file. Another way that we could use linked lists is to superimpose a new structure on top of an existing structure.

Consider a file with some preexisting organization. Consider an attribute $a$ for which each item of interest has some value, say $a_1$. Suppose further that on occasion it is desirable to find all those items for which the attribute $a$ has the value $a_1$. A linked list can provide this function:

- Set up a head for each separate attribute value that $a$ may have.
- There is a head for $a_1$, another for $a_2$, and so forth.
- Provide a field in each record for the attribute $a$.
- In each record, use this field for a pointer instead of for a value.

A pointer appears in the field $a$ regardless if the chain links to the head $a_1$, $a_2$, etc. Now the file is divided into a number of subfiles according to the value of the attribute $a$. To sequence through all records having value $a_i$, belonging to the subfile $a_i$, start at the head assigned to this value. It points to a record for which the corresponding individual has that attribute value. This record points to another record via the *a* pointer. Records point to each other in this fashion forming a trail to one with a terminal pointer, meaning that we have encountered all the records in the subfile (with value $a_i$).

Notice that the order in which we encounter records has no meaning; this order probably corresponds only to the order in which the records were entered into the subfile.

The multilink organization offers two advantages:

- It does not affect the original file organization that it overlays.
- It enables us to retrieve records that have a particular field value without a serial search of the entire file.

It has disadvantages too:

- If only a pointer is found in this field of the record, then the value associated with the field can only be obtained by examining the head or by *another* (additional) field.
- If both a pointer and a value appear in the record, additional space is consumed.
- Updating the file requires updating the linked list also—this means finding the predecessor and successor records for every field for which linked lists are maintained or using the pointer from the head.

### Details

In most instances where a file structure is described, an illustration adds to the comprehension. In this case the relation is so complicated that an illustration only seems to confuse matters. This is because we superimpose two different file organizations on the same list. Let us reexamine a multilink file, considering it to be a set of records already organized in some way. For simplicity, let us consider a sequentially ordered file.

*normal record*     A record is assigned a place in the sequential file according to its key relative to the other records in the file. The record consists of a number of fields. Many of these are normal fields corresponding to attributes of the individual. For each attribute they carry a field value corresponding to the attribute value in the real world. These fields can be used in a normal fashion as described elsewhere.

*linked list*     The records also belong to a number of linked lists. Let us suppose that there are n attributes for which the multilink structure prevails. For each attribute there is a field provided in every record. This field does not contain a value but rather a pointer to another record.

For each attribute, let us suppose that there are a number of distinct values for which we can record that attribute. There is one linked list for each attribute value. Examine Fig. 12.6.1 where three attributes are used to provide the multilink structure—sex, income, and geographical location. In this figure we see there are two alternatives for sex, four for income, and four for geographical location. Hence, there are ten linked lists superimposed on our original sequential structure.

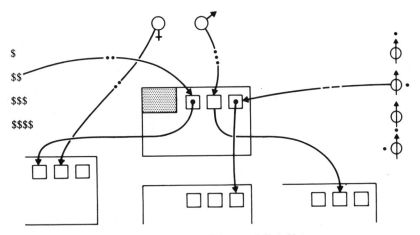

**Fig. 12.6.1.** A record in a multilink file

To describe this symbolically, consider the i'th attribute and suppose that it has the $v_i$ values. Then the number of different linked lists, N, is the sum of these $v_i$'s where i ranges from 1 to $\alpha$, the number of attributes of interest or

$$N = \sum_{i=1}^{\alpha} v_i \qquad (12.6.1)$$

### Linking

For each of $\alpha$ attributes, there are as many heads N as there are attribute values above. This is illustrated with respect to a single record in Fig. 12.6.1. In all but the most complicated cases each record belongs to exactly one linked list with regard to each attribute or a total of $\alpha$ linked lists. Thus for sex there are two linked lists: male and female. The record in the figure belongs to a male. It is on the linked list stemming from the "male" head. There are many other "male" records appearing provided on this linked list. One of them points to this record. The sex field in this record has a pointer to another "male" record. All records are either on the "male" or "female" list.

This record also belongs to someone in the second-income level. The record is pointed to by another record belonging to an individual in the second-income level. There is a pointer in the second link field to a record for an individual in the second-income level, and so forth. If you follow the pointer from the male head, you find all the records that represent males. This feature is obviously useful when collecting information with regard to a particular attribute.

### Updating

When an individual is added to the universe, a record for him is added to the file. Then all the multilinks are updated—one head for each attribute. The original record should contain information about the multilink attributes. The pertinent head is found and repointed to the new record for each value of each link attribute, which in turn is pointed to the record last pointed to by the head. For this technique, heads point to most recent records first.

### Summary

The multilink list is complicated and troublesome but can be an effective way to do certain "planned for" searches.

The multilist finds frequent application and can save valuable processing time when a file lends itself to vertical partitioning.

If the file designer chooses a complex file, you can be sure that the program problem will be complex too. When the file is simple, like a sequential one, the Access Method will do most of the work. Time and space are sacrificed. For a complex file, design and programming take an initial cost toll. In the end, there is a saving in space, IO time, and processing time. Since this is repeated every time the file is used, it may provide an attractive overall saving.

### PROBLEMS

12.1   For the blocked linked list:
    (a) Is it usually ordered—locally or globally?
    (b) What is the sublist?
    (c) Is the sublist dense, loose, or semidense?
    (d) Can one append to a sublist?
        (i) How?
        (ii) What provision is required previously?

(iii) Is this an advantage? Explain.

(e) when a sublist is "full," how is a record appended while maintaining order?

12.2 (a) What is the difference between vertical and horizontal partitioning?

(d) Describe one list structure that uses each.

**For the Multilist:**

12.3 (a) What is the advantage of dividing a file into a header and trailer?

(b) For each record, what determines where fields go into
   (i) the header?
   (ii) the trailer?

(c) Are the header and trailer really files or subfiles? Explain.

12.4 Explain how one finds a record in
(a) the header,
(b) the trailer.

12.5 For the one-to-one header-trailer file (HTF):
(a) Explain with a figure.
(b) What is the general need of the structure?
(c) Give an example not in the text where you might choose this structure.

12.6 For the many-to-one HTF:
(a) Explain with a figure.
(b) What is the general need of the structure?
(c) Give an example not in the text where you might choose this structure.

12.7. For the one-to-many HTF:
(a) Explain with a figure.
(b) What is the general need of the structure?
(c) Give an example not in the text where you might choose this structure.

12.8 For the many-to-many HTF:
(a) Explain with a figure.
(b) What is the general need of the structure?
(c) Give an example not in the text where you might choose this structure.

12.9 For the inventory example in Fig. 12.2.1.
(a) What is the basis for segregating (sub)records into the files **s**, **d**, and **v**?
(b) How does one do a search of the single file **s**, **d**, or **v**?
(c) What pointers exist between files and why?

12.10 Explain maintaining of multilists. Describe how to append or delete header, trailer, or complete record from a
    (a) one-to-one HTF,
    (b) one-to-many HTF,
    (c) many-to-one HTF,
    (d) many-to-many HTF.

**For the Compound File:**

12.11 Examine the structure of the sequential compound file (SCF):
    (a) What is the relation of keys of successive records?
    (b) How do major and minor records differ in function?
    (c) Need the records differ physically? How or why?

12.12 For the SCF:
    (a) Why is it artificial to consider associated records as subrecords?
    (b) How could you restructure it as a HTF? What would the file relation be, one-to-one, etc.?

Consider a checking account application. The master file consists of records with these fields:

| | |
|---|---|
| 1: | account number |
| 2–8: | personal information: name, address, etc. |
| 9–14: | account information: credit rating, charge procedure, statement frequency, balance, deposit activity, check activity, etc. |
| last: | credits and debits |

12.13 Lay out a SCF, **m**.
    (a) Describe the major record.
    (b) Describe the minor record.

12.14 For **m** of Problem 12.13, describe a transaction file **t** that would convey the day's transactions.
    (a) What is the record format and why?
    (b) Suppose that transactions arrive randomly, how is **t** preprocessed?
    (c) Describe posting with **t** and **m**.

12.15 For the file **m** of Problem 12.13, examine the monthly statement procedure.
    (a) How is processing performed?
    (b) What is printed and where on the statement form?
    (c) What happens to **m** after—how is it initialized?

12.16 Describe maintenance for **m**—append and delete.
    (a) How often is it done?
    (b) Describe each input record.

(c) How is it done?

12.17 Now consider keeping **m** as a HTF consisting of $m_H$ and $m_T$.
   (a) What fields of **m** are put into $m_H$ and $m_T$?
   (b) How are $m_H$ and $m_T$ organized?
   (c) Are records in $m_H$ and $m_T$ fixed or variable in size? Explain.

12.18 For **t** of Problem 12.14 and **m** of Problem 12.17, describe daily posting.
   (a) Why must records in $m_H$ be updated during each posting period?
   (b) How is posting done?
   (c) How are $m_H$ and $m_T$ effected?

12.19 Repeat Problem 12.15 for **m** of Problem 12.17.

12.20 Repeat Problem 12.16 for **m** of Problem 12.17.

12.21 Reorganize **m** again as a HTF of three parts to consist of $m_S$ with fixed records containing statement information, $m_P$ with variable records containing personal information, and $m_A$ with a variable number of fixed records containing activity information—credits and debits. Describe posting of **t** for **m** above.

12.22 Describe statement preparation for **m** above.

12.23 Describe maintenance for **m** above.

12.24 Discuss using a linked list trailer for
   (a) $m_T$ of Problem 12.17;
   (b) $m_A$ of Problem 12.21.

12.25 Discuss the pros and cons of applying a linked list format for records of a compound file. Make a sketch of how this would work.

12.26 When **m** for the checking problem is kept on DASD's, as a SCF or HTF, examine
   (a) how the file is broken into subfiles by cylinder,
   (b) how to apply intermediate overflow.

12.27 Describe the multilink list (MLL) in your own terms.
   (a) What is its advantage for search?
   (b) What is its disadvantage for random look up?

12.28 Set up a typical record for a personnel file allocating all required fields including i.d., name, skills, number of dependents, and state. Describe how to convert this into a MLL file.

12.29 Describe in detail how to perform a multiple factor look up in an MLL file, for example, for a particular skill, salary, number of dependents, and state.

12.30 Describe how to add a record to the personnel MLL file in Problem 12.29.

# Table of Operators, Special Symbols, and Observed Words

| Operator | Meaning | Example | Page |
|---|---|---|---|
| $\equiv$ | represented as | $A \equiv Cl$: A is represented as hex Cl | 15 |
| ( ) | contents of | $a = (A)$: contents of cell A is a | 31 |
| [ ] | address of | $A = [a]$: address of record a is A | 31 |
| $\longrightarrow$ | move | $b \longrightarrow D$: move record b to cell D | 31 |
| $\prec$ | precedes | $r \prec s$: record r precedes record s | 34 |
| *num* | number of | *num* $L = 50$: there are 50 cells in L | 53 |
| $\Lambda$ | empty set | $L = \Lambda$: L has no cells | 53 |
| @ | empty flag | $@ \longrightarrow A$: flag cell A as empty | 55 |
| *len* | length | *len* $a = 75$: record a is 75 bytes long | 78 |
| *esearch* | search equal | t *esearch* $L = R$: a record with key the same as t is found in L at R | 92 |
| $\hat{L}$ | average number of looks | $\hat{L} = num\ L/2$: looks to find a record in list L is half the list length on the average | 92 |
| $\infty$ | largest possible key | $\infty \longrightarrow$ (IPR): put the largest possible key into IPR | 117 |
| *stop* | stop | $i \geq j$:  *stop*: stop if $i \geq j$ | 118 |
| $\lceil \rceil$ | ceiling | $n = \lceil \log_2 N \rceil$: n is next integer larger than $\log_2 N$ or $\log_2 N$ if it is integral | 141 |
| $\lfloor \rfloor$ | floor | $S = \lfloor N/2 \rfloor$: S is N/2 for N even and $N/2 - \frac{1}{2}$ for N odd | 147 |
| *point* | forward pointer | *point* $ABE = AL$: ABE points to AL<br>*point* **file** $= ABE$: the head of **file** points to ABE | 163 |
| *egsearch* | search equal, greater | bob *egsearch* file $= CAL$: (CAL) bob where CAL is a cell for file | 163 |
| *back* | backward pointer | *back* $E = D$: E points back to D | 178 |

| Operator | Meaning | Example | Page |
|---|---|---|---|
| * | terminal pointer | *point*$^n$ **file** $=$ *: the nth cell of **file** is last | 181 |
| *branch* | branch pointer | *branch* B $=$ X: B has a "branch" pointer to X | 188 |
| *map* | mapping function | *map* k $=$ $L_i$: the key k maps into cell $L_i$ | 243 |
| *go to* | go to | i $+$ 1 $\longrightarrow$ i; *go to* 3: set i to i $+$ 1 and return to step 3 | 256 |
| ¢ | delete flag | ¢ $\longrightarrow$ R: indicate in cell R that the record held there is now deleted | 260 |
| *probe* | probe algorithm | *probe*$^3$ *map* k: map k into a neighborhood and then apply the probe algorithm three times | 271 |
| *height* | height | *height* $Tt_1 = h + 2$: the height of the subtree made of the subtree $t_1$ and vertex T is h $+$ 2 | 327 |
| $\cup$ | concaternate | $n' = n \cup d'$: the record n' is found by pasting together the record n and the record d' | 354 |

# APPENDIX B

# *Glossary*

**binary tree total directory file**   a binary search tree where each cell     311
instead of containing a datum contains a pointer thereto.

**birth**   the creation of a new record in a file for a new individual who has     5
joined the universe.

**bit**   an atom of information in the form of 1/0, on/off, or yes/no.     13

**bit site**   the place on the medium where a single bit is or may be recorded.     19

**block**   the amount of information read from or written onto the medium     19
by a DEVICE in a single request.

**blocking**   the action of placing several records into one block.     26

BUFFER   a hardware DEVICE component that stores information as it     18
passes between the DEVICE and the CHANNEL.

**buffer**   an area in MEMORY that provides intermediate storage for a block     26
of information.

**byte**   eight bits which represents a symbol of the alphabet.     14

**cell**   the place where a record is stored.     22

**cell, branch**   a cell in a linked list that can be used to point to at least     185
two other cells.

**cell, branch-only**   a branch cell that does not contain data.     188

**cell, branch-only, bifurcating**   a branch-only cell that contains exactly     188
two pointers.

**chain**   see *arc sequence*.     39

CHANNEL   hardware that enables the transfer of data between a DEVICE     8
and MEMORY without interfering with processing.

**channel program**   a set of instructions passed by the CHANNEL to the     26
DEVICE, causing the DEVICE to perform a series of activities.

**character**   a symbol, the basic atom for constructing information at the     5
user level.

**character set**   that set of characters available to the user in this system.     5

**check digit**   a digit, arrived at by performing an algorithm, that is     97
appended to a record identifier for the purpose of validating keyboard
entry.

**collating sequence**   see *order relation*.

**compact**   to rewrite a file so that holes found in the list containing that     168
file now become available as cells that may contain records to be appended to the file.

CONTROL   the subsystem that directs the COMPUTER to process data per     8
the instructions found in the program.

**cycle**   an edge sequence for which the origin and destination is the same     38
(see *elementary*, *simple*, and *nonsimple*).

**data**   representations of the real world in a form that can be stored in     1
the COMPUTER.

**death**   the destruction or deletion of a record from the file because an     5
individual corresponding to that record has left the universe.

**deblocking**   the action of the Access Method in retrieving a record from   26
a block for the user.

**deck**   a volume of punchcards.   20

**dense**   a list is dense with respect to a file when each of its cells contains   53
a record of the file.

DEVICE   hardware that enters or removes information from a medium.   17

**digraph** (directed graph)   a graph containing arcs instead of edges.   37

**directory**   an auxiliary list which contains entries which enable us to find   214
a record, a sublist containing that record, or a subdirectory which will
take us eventually to a neighborhood containing the record.

**doubly linked branch list**   a doubly linked list that also contains branch   203
cells.

**doubly linked list**   a linked list where each cell contains a forward and   177
reverse pointer.

**edge**   a line segment without direction.   35

**edge sequence**   a sequence of edges providing a continuous trip from   38
an initial vertex to a terminal vertex (see *elementary*, *simple*, and *non-simple*).

**element**   the place where a field resides.   23

**elementary**   when applied to an edge sequence, cycle, etc., indicates that   38
no edge or vertex appears more than once.

**empty**   a list is empty when no cell contains a record.   53

**fence**   the cell that divides a list or sublist into two approximately equal   139
parts.

**field**   the place in a record that stores information about a single   3
attribute.

**field, fixed**   one that is the same length for every record in the file and   59
has the same relative position in each record.

**field name**   a name used to distinguish one field in a record from another   4
and generally corresponding to the attribute name.

**field value**   the value contained in a field, corresponding to the attribute   4
value of an individual in our universe.

**FIFO**   see *queue*.

**file**   a collection of records, one for each individual in our universe.   3

**file, activity**   the set of records in the master file for which activity has   95
taken place during a given period.

**file, complex**   a file where several relations exist between the records of the   347
file simultaneously.

**file, directory**   a file using a directory where the directory points to   222
a sublist where the record sought will be found if present.

**file, directory, two-level**   a file using a directory, where each entry points   226
to a subdirectory and the subdirectory points to a subsublist where the
record may be found.

**incidence**   the number of edges or arcs which terminate at the given      42
vertex.

**individual**   an object of interest for the data processing problem.      2

**Input/Output Block** (IOB)   a control block which monitors the filling      27
or emptying of a buffer.

**Input/Output Supervisor** (IOS)   the software that receives requests from      25
the Access Method and gives hardware input and output requests to the
CHANNEL.

**intransitive**   a relation that is not transitive.      35

**justify**   to place a value of smaller size than the field into that field so      61
that the value lines up with one side of the field.

**key, fence**   the key of the record contained in the fence.      139

**key** (key field)   a field in the record that uniquely identifies that record.      32

**lateral**   perpendicular to the direction of motion of the medium through      19
the DEVICE.

**leaf**   a pendant vertex that has no arcs leaving.      42

**legitimate** (record or key)   where a key follows a rule for determining      245
if its structure is proper.

**level** (of a leaf or node)   the number of arcs intervening between the      43
root and the given vertex.

**library**   a collection of files representing several universes with respect to      6
a given application.

**LIFO**   see *stack*.

**list**   the place where a file is stored.      23

**list, fixed format**   a list with accessibility limitations so that only certain      169
cells of the list can be accessed.

**list, head**   a set of heads mapped into by a hashing function for the      298
hashed head multiple linked list file.

**list, linked**   a list which is generally not an ordered list but contains a      159
file which is ordered by dint of pointers contained in each record.

**list, ordered**   a list is ordered when records contained in successive cells      56
have keys whose relation is described by the collating sequence.

**list, pushdown**   see *stack*.

**list, random**   see *list, sequential*.

**list, sequential**   a list where the records contained in successive cells are      58
determined by the order in which records were presented in the input
stream.

**list, space**   a linked list of empty available cells.      173

**list, space, branch**   a space list containing empty branch cells.      189

**longitudinal**   in the direction of movement of the medium as it passes      19
through the DEVICE.

**loop**   an arc sequence whose origin and destination are the same (see      39
*elementary*, *simple*, and *nonsimple*).

**loose**   a list is loose when it is neither dense nor empty.   54

**maintenance, file**   adding or deleting records from the file to reflect   6
changes in the universe because individuals have been added or subtracted
therefrom.

**many-to-one**   a mapping between two subsets where several members   246
of the first correspond to only one member of the second.

**mapping**   an operation, arithmetic or otherwise, upon a sought key to   243
yield the address of a neighborhood wherein the record should be found.

**mathematical model**   a representation of the real world using mathe-   1
matical formulas so that a given value of variables can be used to predict
a future state of the world.

**medium**   a physical configuration used to store information.   17

**memorize**   to place a datum into the MEMORY.   10

MEMORY   the subsystem that holds data.   9

**merge**   to combine two or more files of ordered subfiles into a single   123
file of ordered subfiles.

**miss**   when mapping takes us to the wrong record for retrieval or   244
deletion or to an occupied cell for appending.

**multifield**   a field consisting of multiple subfields.   68

**multiprogramming**   where several programs occupy MEMORY simultane-   25
ously and the operating system may switch from one to the other very
quickly.

**neighborhood locatability**   a technique or list structure which enables   213
us to find a neighborhood for further search which will contain a sought
record if present.

**nibble**   half a byte or four bits.   15

**node**   a vertex that is incident to at least two edges.   42

**node, leaflike**   has more arcs entering than leaving.   42

**node, linking**   has as many arcs entering as leaving.   42

**node, rootlike**   has more arcs leaving than entering.   42

**node, simple**   has one arc entering and one leaving.   43

**nonsimple**   when applied to an edge sequence, cycle, etc., indicates that   39
it is neither simple nor elementary.

**occupancy ratio**   ratio of cells in a list that contain records to the list   53
size.

**operand**   a datum to be worked upon by the PROCESSOR.   7

**order, ascending**   an ordered list is in ascending order when successive   56
cells contain records, the key of each of which is higher than that of
the record in the previous cell.

**order, descending**   an ordered list is in descending order when records   56
in successive cells have keys that diminish in magnitude.

**order relation**   a relation that exists among the key fields so that the   32
records containing them can be placed in a corresponding order.

**reel**   a volume for a tape drive.     20

**reflexive**   a relation that applies between an object and itself.     34

**relocate**   for a mapping, where a base quantity is added so that the     249
resulting area is relocated in MEMORY above existing software.

**result**   a new datum produced by the PROCESSOR.     7

**retrieval**   reference to a record of a file without making alteration of that     87
record.

**rollout**   during a merge, when one or more input files has become     128
exhausted.

**root**   a pendant vertex into which no arc enters.     42

**route**   see *edge sequence.*

**scaling**   where a mapping is multiplied by a constant to increase the size     250
of the resulting neighborhood.

**search, binary**   an efficient means for searching an ordered list by divid-     138
ing it into two approximately equal parts recursively.

**semidense**   applied to a list that contains a dense and an empty part.     107

**simple**   when applied to an edge sequence, cycle, etc., indicates that no     39
edge appears more than once.

**simultaneity**   see *overlap.*

**sort**   to create an ordered list from a disordered file or list.     112

**stack**   a list where records are entered and retrieved from the same end     169
so that the last one entered will always be the first one removed.

**stack, space**   see *list, space.*

**step-down**   during a merge, the action of reaching the end of an ordered     125
subfile on one of the input files.

**subfield**   a reasonable subdivision of a field.     64

**subfield, length**   provided to indicate the length of a value subfield that     65
follows or is found elsewhere.

**sublist, logical**   a sublist defined by the physical boundaries together     282
with all other records in overflow assigned to the same sublist.

**sublist, physical**   a sublist as defined by the physical boundaries thereof.     282

**sublist, shared**   in a branching linked list, a sublist that is pointed to by     207
two or more cells.

**sublist, target**   a neighborhood reached by mapping.     243

**subrecord**   a collection of disparate fields with some meaningful asso-     74
ciation.

**subset**   with regard to a set, a collection containing some of the members     6
of that set.

**symmetric**   a relation that if it applies to two elements in one order,     34
also applies to those same elements when their order is reversed.

**table of contents**   see *directory.*

**total directory system**   a file that uses a directory where each directory     215
entry points to a single record and there is one entry for each record
iu the file.

**track**   a longitudinal frame.     19

TRANSDUCER   the active part of the DEVICE that changes electrical signals     18
into deformation of the medium or detects deformation of the medium,
converting it into electrical signals.

**transitive**   if the relation R holds between a and b *and* it holds between     35
b and c, then it must hold between a and c.

**tree**   a graph that contains no cycles.     41

**tree, balanced**   a binary tree where the height of the upper subtree at a     321
node never differs by more than one from the height of the lower subtree.

**tree, multiway**   where each node can point in three or more directions.     332

**tree** (optimum)   a tree for which all branches have a height of either h     321
or h − 1.

**tree, ternary**   where each node is capable of branching in three direc-     330
tions.

**truncate**   where a value is larger than the field provided for it, part of     61
the value is discarded.

**truncation, suffix**   a mapping where the last part of the key is discarded.     247

**trunk**   for a branching linked list, those cells that are reached by     192
following the forward pointer from the head to the terminal cell.

**universe**   collection of objects that might be eligible to be considered     2
as data for a given problem.

**update**   alter a record to reflect changes in the attributes of the     87
individual it represents.

**vertex, pendant**   one that is incident to only one edge.     42

**volume**   the quantity of data that can be mounted upon the DEVICE.     20

**way** (of a merge)   the number of input files to be merged.     128

**well-defined**   a mapping is well-defined when the neighborhood into     247
which it maps is always known.

**write**   to enter information onto a medium.     17

# Index